ECOTEXTILES

The Textile Institute and Woodhead Publishing

The Textile Institute is a unique organisation in textiles, clothing and footwear. Incorporated in England by a Royal Charter granted in 1925, the Institute has individual and corporate members in over 90 countries. The aim of the Institute is to facilitate learning, recognise achievement, reward excellence and disseminate information within the global textiles, clothing and footwear industries.

Historically, The Textile Institute has published books of interest to its members and the textile industry. To maintain this policy, the Institute has entered into partnership with Woodhead Publishing Limited to ensure that Institute members and the textile industry continue to have access to high calibre titles on textile science and technology.

Most Woodhead titles on textiles are now published in collaboration with The Textile Institute. Through this arrangement, the Institute provides an Editorial Board which advises Woodhead on appropriate titles for future publication and suggests possible editors and authors for these books. Each book published under this arrangement carries the Institute's logo.

Woodhead books published in collaboration with The Textile Institute are offered to Textile Institute members at a substantial discount. These books, together with those published by The Textile Institute that are still in print, are offered on the Woodhead web site at: www.woodheadpublishing.com. Textile Institute books still in print are also available directly from the Institute's website at: www.textileinstitutebooks.com.

Woodhead Publishing also publishes the two Textile Institute journals: *The Journal of The Textile Institute* and *Textile Progress*. Both journals are editorially independent of the publisher, and have their own editors-in-chief and editorial boards. Details of these journals and how to subscribe to them can be found at: www.woodheadpublishing.com.

Ecotextiles
The way forward for sustainable development in textiles

EDITED BY
M. Miraftab and A. R. Horrocks

CRC Press
Boca Raton Boston New York Washington, DC

WOODHEAD PUBLISHING LIMITED
Cambridge England

Published by Woodhead Publishing Limited in association with The Textile Institute
Woodhead Publishing Limited, Abington Hall, Abington
Cambridge CB21 6AH, England
www.woodheadpublishing.com

Published in North America by CRC Press LLC, 6000 Broken Sound Parkway, NW,
Suite 300, Boca Raton, FL 33487, USA

First published 2007, Woodhead Publishing Limited and CRC Press LLC
© 2007, Woodhead Publishing Limited, except Supply chain partnerships for sustainable textile
production © 2004, John Easton
The authors have asserted their moral rights.

British Library Cataloguing in Publication Data
A catalogue record for this book is available from the British Library

Library of Congress Cataloging in Publication Data
A catalog record for this book is available from the Library of Congress

Woodhead Publishing ISBN-13: 978-1-84569-214-8 (book)
Woodhead Publishing ISBN-10: 1-84569-214-4 (book)
Woodhead Publishing ISBN-13: 978-1-84569-303-9 (e-book)
Woodhead Publishing ISBN-10: 1-84569-303-5 (e-book)
CRC Press ISBN-13: 978-1-4200-4444-7
CRC Press ISBN-10: 1-4200-4444-3
CRC Press order number: WP4444

Printed by Antony Rowe, Wiltshire, England

CONTENTS

INTRODUCTION

Mohsen Miraftab and Richard Horrocks
Centre for Materials Research and Innovation, University of Bolton, UK

In the nineteenth and twentieth centuries where the majority of industrial development and advancements in science and technology took place, the emphasis had been by and large on meeting demands, reaching manufacturing efficiencies and competing for markets. The notion of pollution, waste and the state of the environment either did not enter this great drive for industrialisation or if it did it took a very low priority indeed. This rather simplistic and maybe convenient approach treated natural resources as if there would be infinite supply and assumed somewhat magical regeneration of resources with no detrimental consequences.

However, during the late 20th century, development in computer and related technologies have created communication networks, which coupled with faster and more economic international travel have played major roles in creating the concept of the "global village" where the general consuming public across the world are much better informed of the harm and the dire effects of what man is doing to himself. Global warming, melting of glaciers, increasing numbers and severity of floods and hurricanes, appearances of new diseases and associated viruses, despite medical advances, are all alarming signs of what is happening to our environment.

To be able to continue global economic development in the present century, issues of sustainability, renewability and recyclabilty must become central to all wealth-creating activities. The slogan of "polluter pays" must be engraved deep in any governmental and private activity to ensure full commitment to environmental concerns. A unified and all-inclusive set of actions need to be taken not only to clean up the environment but to develop new and novel methodologies to at best eliminate waste and at worst minimise their creation. For such actions to occur requires international co-operation and agreement at all levels of waste productions, its minimisation and not least its effect on trade.

The much talked about Kyoto Treaty, instigated in 1997, was the first serious attempt by industrialised nations to curb emission of greenhouse gases obliging them to reduce emissions by around 5.2% below their 1990 levels over the following decade. However, the USA responsible for at least 55% of world's toxic gas emissions would not ratify the Treaty. The revised version made considerable compromises to satisfy all concerned and cut the 5.2% reduction requirement to 2% but still failed to win the support of the US before coming into force in February 2005.

Acting independently, the European Union over the past decade or so has tried to enforce a number of directives through its member states with the intention of monitoring and controlling harmful gases and hazardous waste feeding into the environment. The first of these has been the Packaging and Packaging Waste Directive (94/62/EC) implemented in the UK through the Producer Responsibility Obligations (Packaging Waste) Regulations in 1997. According to this directive, businesses with a turnover above £5 million and handling over 50 tonnes of packaging per year have to register with the Environment Agency or a compliance scheme. They also need to supply annual packaging flow information to prove they are meeting their recycling and recovery obligations.

The European Landfill Directive followed next and had to be implemented in member countries by 2001. The Landfill Directive demands reduction in landfilling of biodegradable municipal waste to 75% of 1995 levels by 2006, 50% by 2009 and 35%

by 2016. Countries with major reliance on landfill, such as the UK, will be allowed to claim exemption to delay meeting the targets till 2020. The directive also forbids co-disposal of hazardous and non-hazardous wastes and restricts landfilling of liquid wastes, clinical wastes and other similar materials. Under this directive landfilling of whole and shredded tyres are also banned from 2003 and 2006 respectively.

The proceeding directive (i.e. 2000/76/EC) deals with the issue of incineration and had to be implemented by December 2002. This directive aims to reduce emissions to air, water and land from the incineration of non-hazardous and hazardous wastes. The directive applies to a range of municipal waste, sewage sludge and clinical waste incinerators, as well as a variety of incinerators burning treated waste wood and waste oil. As a consequence, all incinerators in the UK have had to meet the tighter emission requirements of this directive by December 2002.

The waste from Electrical and Electronic Equipment (WEEE) Directive was proposed in June 2000 and, following long delays, it is to be implemented in summer of 2006. This directive provides guidelines for collection, disposal and recycle/re-use of large and small domestic appliances, electrical and electronic tools, toys, monitoring instruments, automatic dispensers as well as computers, lighting and medical equipment. The directive also calls for phasing out the use of heavy metals such as lead, cadmium, mercury and hexavalent chromium, and certain halogenated flame retardants - polybrominated biphenyls (PBBs) and certain (principally the penta- and octabromo-variants) polybrominated diphenyl ethers (PBDE) by 1st January 2008.

The End-of-Life Vehicles Directive (2000/53/EC), one of the latest Directives to be implemented, concerns cars, vans and certain three-wheeled vehicles. Its main requirements are to limit the use of certain hazardous materials in the manufacture of new vehicles and automotive components and to promote the recyclability of vehicles. End-of-life vehicles need to be de-polluted prior to dismantling, recycling or disposal. Producers will be required to pay full or substantial part of the cost of these recoveries by January 2015.

Policing implementation and ensuring full compliance with these directives is a huge and a daunting challenge and may take many years to take roots and become fully established. But given increasing public concern and the apparent readiness of communities around the globe to meet the associated costs and "inconveniences", it will only be a matter of time before it becomes a universal culture and an ethical measure in that societies respect the environment in which we all live.

As yet there has not been a European directive specifically targeting textiles and textile-related material waste. However, given the volume of synthetic and organic polymers used for various applications, it will not be long before such directives are instigated and demand for their implementations is made. Some carpet manufacturers including Honeywell, BASF and Du Pont already run schemes whereby old carpets are exchanged for new carpets with some degree of success. This kind of marketing strategy could very well be preparing the grounds for imminent directives concerning all textile-related products.

It is in light of these inevitabilities that the textile manufacturers directly or indirectly are increasingly sensing the need to act and clean up their industry. This will only be achieved by serious investment and sustained commitment to new and novel technologies that would eliminate/minimise waste generation and reduce or de-toxify emitted gases and effluent by-products. While national and international regulations will drive this waste minimisation need forward, we must not forget that challenges remain at technological, fashion and design levels to ensure that recycled textiles and waste products may appear once again in consumer products using processes that are also commercially viable and successful.

In order to play a part in informing world's industrial and academic communities of the scientific and technological advances made in the eco-friendly textile sector, the University of Bolton since 1995 has organised and hosted a series of international conferences on the ecological aspects of textile or "ecotextiles". The prime aims of these conferences have first and foremost been to bring together like-minded specialists across the textile manufacturing, fashion and design spectrum and from around the world and to provide an international platform for exchange of ideas and examination/assessment of innovative methodologies/approaches in tackling the menace of waste and pollution.

The first Ecotextile conference took place in April 1995 and adopted the title "Wealth from Waste in Textiles" as its theme. The conference attracted over eighty delegates from UK and many countries overseas. Over 25 papers and posters were presented and the proceedings were published by Woodhead Publishing. The second conference was held in April 1998 under the following title "Sustained Development"; this conference attracted an even bigger audience and received sponsorships from no less than six industrial and research organizations. Over 120 attended this conference and 36 papers and posters were presented by various delegates from around the world. The proceeding from this conference was also published in 2000 by Woodhead Publishing. This latest Ecotextile conference bearing the following theme "The Way forward for Sustainable Development in Textiles" was held in July 2004 and over 35 international speakers delivered many practical and futuristic solutions in their presentations. A selection of 23 papers presented at this conference have been carefully edited and categorised under different headings for publication in this book.

In editing the papers in this volume, we have tried to maintain the individuality of style of each presentation while assuring an overall requirement of uniform formatting. The validity of technical contents and discussions are the sole responsibilities of the respective authors and readers should focus their enquiries direct to them should the need arise.

Part I

Recycling and use of waste as raw materials

PERCEPTIONS TOWARDS CLOTHES WITH RECYCLED CONTENT AND ENVIRONMENTAL AWARENESS: THE DEVELOPMENT OF END MARKETS

Yukie Nakano

Centre for Design Research, Northumbria University, Sandyford Road, Squires Building, Newcastle Upon Tyne, UK.

ABSTRACT

UK waste is growing steadily every year by 2-3 per cent, despite the fact that our government encourages us to recycle more. Due to a lack of landfill sites, we need to reduce our waste drastically. However, developing end markets for recycled materials is necessary for recycling to be successful, and the achievement of those markets depends on consumer demand for products.

This study has explored the public perception towards clothes with recycled content and was designed as a guide to clothing industries that are, or will be, dealing with recycled materials for their products. The public survey aimed to identify the potential market for clothes with recycled content, particularly those made from plastic (PET) bottles using a fleece jacket as an example.

The findings reveal that there is a contradiction between the public reaction towards products with recycled content and their awareness of environmental issues.

INTRODUCTION

Pressure to meet EU landfill regulations, a lack of available landfill sites and a growing concern for environmental hazards such as land, water contamination and air pollution associated with incineration and landfill sites have pressurised the UK Government to take action and raise the recycling rate. They indicated a target in Waste Strategy 2000 for England and Wales to recycle 25 per cent of household waste by 2005.[1]

However, in 2001/02, we created 28.8 million tonnes of municipal waste of which 22.3 million tonnes went into landfill. This was an increase of 2.4 per cent compared to 28.1 million tonnes in 2000/01 with 22.1 million tonnes going into landfill. Parallel to this development there was also a 1.2 per cent increase in recycling municipal waste from 12.3 per cent in 2000/01 to 13.5 per cent in 2001/02.[2] This clearly shows that although the recycling rate has seen a rise, this is not sufficient to reduce the amount of waste going into landfill sites, as the total amount of waste produced has increased at a faster rate than recycling. If our pattern of consumption keeps to its current trend and we do not reduce our waste, it is clear that in the near future there will be no available landfill sites.[3]

One significant contributor to this situation is the plastic beverage bottle. It is derived from crude oil and the rapid growth of plastic bottle consumption combined with its short life cycle, results in its continuous entry as waste into landfill sites and incinerators.[4] Looking at waste by weight, approximately 10-11 per cent of household wastes is plastic.[5] Around 1.6 per cent of this comprises plastic bottles. However, looking at waste by volume, plastic bottles would represent more than 5.4 per cent of the waste.[6]

3

In England recycling plastic started in the 1990s, since then the development of plastic recycling has been slow: approximately 3% (14,000 tonnes) in 2001.[7]

The main key barriers to the successful implementation of plastic bottle recycling have been identified as the lack of an efficient recycling infrastructure and an unstable market for collected materials.[8] While improving the recycling infrastructure is a relatively straightforward task, which some local authorities may already have started addressing, developing end markets for recycled materials provide bigger challenges.

As Sutherland mentions, developing end markets for recycled material is vital for recycling to be able to grow.[9] He also suggested that success of the market development relies on consumer demand for goods with recycled content.[10]

The RECOUP survey shows that approximately 10% of local authorities are not collecting plastic bottles because they have no confidence that there is a market for the collected materials.[11] The recent WRAP study also pointed out that the plastic industry's preconceptions towards recycled plastic materials have been hindering market development. The industry believes that recycled materials are of inferior quality to virgin materials and therefore hesitate to use them.[12] Nevertheless, advances in recycling technology have made it possible to produce desirable quality goods from recycled plastic. In order to encourage industries to change the current practice it is necessary to convince them of the quality of recycled materials and to examine the public acceptance towards products with recycled content.

This study concentrates on clothing with recycled content, particularly those made from plastic (PET) bottles. In order to examine the public perception towards clothes with recycled content, a survey was carried out using a fleece jacket as an example. The fleece jacket is a product that can be made from recycled materials and it is a particularly good example of recycled clothes currently available in the market place as it shows that recycled materials are no longer inferior quality to virgin materials.

The survey aimed to identify the potential market for clothes with recycled content. This study and the resulting information was designed as a guide to clothing industries that are or will be dealing with recycled materials for their products. Shopping behaviour and environmental awareness, especially attitudes towards recycling were closely scrutinised. Analysis of previous consumer attitudes[13] towards products marketed as environmentally friendly have contributed to the design and the analysis of the survey.

BRIEF HISTORY OF THE GREEN MOVEMENT IN THE CLOTHING INDUSTRY

In the early 1990s, we all experienced an increased interest in the green movement. The products e.g. biodegradable washing detergent, organic cotton and paper made from controlled forests, marketed as 'environmentally friendly' were sold everywhere. Yet, consumers were soon disappointed with lesser quality and the premium price of its products compared to conventional ones.

The clothing sector tells a similar tale. Esprit launched 'Ecollection' using mainly organic cotton with natural colours. This was initially an overwhelming success but soon consumers moved to more colourful clothes.[14] Patagonia invented a fleece jacket made from recycled PET plastic bottles. The product was very successful and many companies followed suit and produced a fleece jacket made from plastic bottles. However both the jacket's popularity and the high price gave rise to cheaper copies by non-high market clothing manufacturers that used virgin materials. As a result, there are plenty of low priced fleeces in the market place but they are not made from recycled materials.[15]

The green movement in the early 90s was a fashion. It arrived suddenly and disappeared the next season, just like other fashion trends. In order to sustain such a movement, it cannot be based on short-term trends, but needs to be linked to real long-term benefit. In the US, organic fibre (cotton) production followed the growth of the organic food movement.[16] It seems the UK followed a similar trend. The recent food scares in the UK triggered many to search for safer food. People have started to look into where products come from and how they are produced. As a result there has been a rapid growth of organic and fair-trade food products and the certified organic cotton and fair trade textile/clothing businesses have recorded sales growth. For instance, the organic baby and toddler brand 'Green Baby' recorded more than £2m in sales in 2003/04 and expects sales to go up to £5m by 2006.[17] The fashion designer Katharine Hamnett has also recognised the opportunity and is planning to launch an ethical fashion business in spring 2005.[18]

It seems that certified products are widely acknowledged and have become desirable products in recent years. The pressing question is whether products with recycled content could follow suit and see increased popularity. Both organic and recycled products have some benefits towards our environment. However, there are fundamental differences between these type of product. Organic fibre has added value compared to conventional cotton fibre. Neither pesticides that are considered toxic nor fertilisers are used during production.[19] So that organic cotton offers assurance that the skin is not in direct contact with chemicals. This represents a significant benefit for those with a sensitive skin, e.g. babies.

Recycled fibre on the other hand does not offer immediate benefits or reassurance. On the contrary, recycled fibre is often associated with having been "used", and is thought of as having as inferior quality to new materials. It is therefore necessary to investigate public perception towards clothes with recycled content in order to identify a potential market for clothes with recycled materials.

METHODOLOGY USED IN THE SURVEY

The public survey was carried out in the city centre of Newcastle-upon-Tyne over a period stretching from the 26th January to 3rd February 2001. The particular location was considered the most suitable, following McCormack and Hill's (1997) suggestions, Eldon Square was chosen as a location for the survey because it is seen as a busy retail centre and is easily accessible to a a cross-section of the population in a relatively short period of time.[20] The data were collected from 95 respondents mainly aged between 19 and 60 years old. Particular attention was paid not to collect data from only certain groups e.g. gender and age. In order to avoid irrelevant respondents, filter questions were asked before filling in the main questionnaire. Respondents in the survey choose outer clothes by themselves and their occupations were not related to fashion, textiles or clothing retail, and are not dealing with environmental issues as part of their occupations. Face-to-face interviews were used to collect data and the interviewer could also answer the respondent's questions. Though efforts were made not to create bias, it may have arisen through the interviewer's sub-conscious non-verbal signals or through misunderstanding of questions and answers.

The sample range

48.4 per cent (n*=46) of the respondents were male and 51.6 per cent (n=49) were female. The respondents fit into age groups of 46 to 60 (28.4 %, n=27), 26 to 35 (18.9 %, n=18), 19 to 25, 36 to 45 (16.8 %, n=16), the age group of under 18 (10.5 %, n=10) and the age group above 61 (8.4 %, n=8). More than 80 per cent of the respondents were aged between 19 to 60 years old. The population of Newcastle-upon-Tyne at the time of the survey was 259,536 and approximately 10 per cent of its population were aged 20 to 24. This could be explained by the concentration of universities and colleges within and around the city. The economically active population in the city was reported to be 58.5%, which is lower than the England average of 66.9%.[21] *(n stands for number of respondents)

The questionnaire was divided into several sections. Firstly filter questions were asked. Then in section one, respondents were asked about their knowledge of, and reaction towards, eco-clothes made from recycled materials using the fleece jacket as an example. A fleece fabric sample made from recycled plastic bottles was used to help the respondents get a realistic idea of what recycled products could look like. Some of the questions were multiple choice questions and dichotomous (yes, no) questions. The option 'do not know' was given with the appropriate questions. These questions were particularly useful as they allow respondents to miss out irrelevant questions.[22]

In section two and three, the respondents were asked about their shopping behaviour in relation to clothes and their attitudes towards products which were marketed as environmentally friendly. In section four, they were asked about their awareness towards environmental issues especially recycling.

Finally, classification questions such as age and residential location were asked. The questionnaire was designed with closed-ended questions using semantic differential scales for most of the questions in order to gain 'the closest to true internal data'.[23] Some of the questions asked to list options e.g. 'Paper' and 'Plastic Bottles', prompting the respondents to show a degree of preference according to a scale from one (not at all) to five ('all the time'). 'Other' was included in the list to allow for further options. This enabled respondents to express themselves by filling in their own option,[24] in case they could not find a suitable option in the list.

The data was analysed using a software package called Statistical Package for the Social Sciences (SPSS). A significance level of 95% (or 0.05) was set.

THE SURVEY FINDINGS

The survey was designed to:
-   identify the public reaction towards clothes with recycled contents using a fleece jacket as an example;
-   examine key drivers when they purchase clothes;
-   find out public awareness of environmental issues especially recycling; and
-   identify the potential market for clothes with recycled content.

The survey results and questions are outlined below:

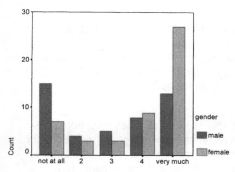

Fig 1 Plastic bottles/gender (n=94)

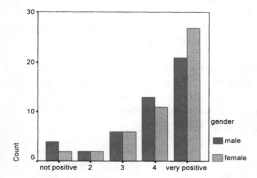

Fig 2 Reaction/Gender (n=94)

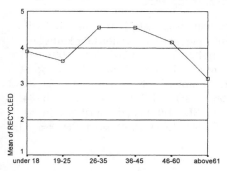

Fig 2.1 Reaction/Age (n=94)
1=Not satisfied, 5=very satisfied

Question 1 - Would it surprise you to know that fleece garments can be made from recycled plastic bottles?

The data is divided into 'not at all surprised' (23.4%, n=22) and 'very much surprised' (42.6%, n=40). It seems that there is a knowledge gap among people. Females are more surprised to know that fleece jackets can be made from recycled plastic bottles than males (Sig* .04) (see Fig 1).
*(Sig stands for significant value)

Question 2 - If you are told that your fleece garment is made from recycled materials, what would be your reaction?

This question was asked in order to investigate the reaction towards recycled materials especially for clothing. 51.1% (n=48) of the respondents chose 'very positive'. 76.6% (n=72) of the respondents chose 'positive' and 'very positive' (see Graph 2). The majority of people seem not to discriminate against recycled materials. Digging deeper, we can see that the age group 26 to 45 could be a target for clothes made from recycled materials. Fig 2.1 clearly shows that the age group '26-35' (mean 4.56) and '36-45' (mean 4.56) have the highest mean values among the age groups implying that they have a positive reaction towards recycled materials more than any other age group. The t-Test results shows that there is significant difference between the age group 'under 25(n=26)' vs '26-45(n=34)' (Sig .002) and '26-45(n=34)' vs 'above 46(n=34)' (Sig .024).

7

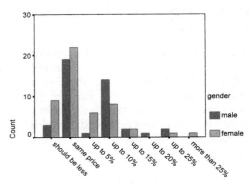

Fig 3 Extra pay/Gender (n=91)

Question 3 - Recycled materials cost more at the moment. How much extra would you be prepared to pay for a garment made from recycled materials compared to new materials?

45.1% (n=41) of the respondents are prepared to pay the same price for recycled materials as for new materials and 31.9% (n=29) are prepared to pay up to 10% more than for new materials. If products made from recycled materials cost more than 10% than products made from new materials, they will not appeal to people (see Fig 3).

89.9% (n=80) of the respondents did not know of any clothing companies, which are dealing with environmental issues. 60.7% (n=54) of the respondents never bought environmentally friendly clothes and 36% (n=32) are not aware of such products. Only 3.4% (n=3) of the respondents actually have bought eco-clothes. Only 9.1% (n=8) said that they consider the environment when they shop for clothes.

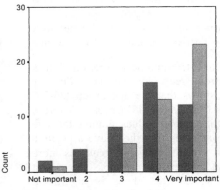

Fig 4 Touch/gender (n=84)
1=Not important 5=Very important

Question 4 - Think about the last knitted item (e.g. jumper, sweater...) you purchased. What influenced your decision?

This question was asked to find out the respondents' purchasing decisions for knitwear. 'colour' (n=82, mean 4.26, SD* .90) and 'Design' (n=81, mean=4.15, SD .91) are top of the list, followed by 'comfort' (n=86, mean=4.06, SD 1.01), 'touch' (n=84, mean=4.06, SD 1.05) and 'quality' (n=81, mean=4.05, SD 1.09). *(SD stands for standard deviation)

'Touch' seems more important to females (n=42, mean 4.36, SD.88) than males (n=42, mean 3.76, SD1.12) (Sig .031) (see Graph 4). Also the data indicates that the age group 36 plus (n=42, mean 4.43, SD.94) is likely to care more about the 'Quality' of knitwear garments than age groups under 36 (n=39, mean 3.64, SD=1.11) (Sig .001) (see Fig 4.1).

26.7% (n =24) of the respondents do not buy particular branded clothes at all. However, the rest of the respondents choose 'quality' (n=69,

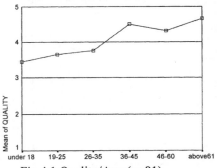

Fig 4.1 Quality/Age (n=81)
1=Not important 5=Very important

mean 4.25, SD.86) as the main reason for them to purchase branded clothes followed by 'design' (n=67, mean 3.96, SD1.11) and 'value for money' (n=68, mean 3.88, SD1.06).

Question 5 - Do you buy environmentally friendly products? (e.g. washing detergent, toilet paper, car etc)

60.5% (n=52) of the respondents have bought products that were marketed as environmentally friendly and 23.3% (n=20) of the respondents said that they have never bought any environmentally friendly products. 16.3% (n=14) of the respondents are not sure if they have or not.

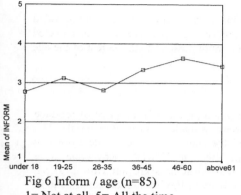

Fig 6 Inform / age (n=85)
1= Not at all, 5= All the time

Question 6 - Do you try to inform yourself about environmental issues, or not?

The data did not indicate a significant gender difference. However, there is a difference among the age groups. Fig 6 shows that the '36 and over' age group (n=44, mean 3.50, SD 1.15) seem to be trying to inform themselves more about environmental issues than 'younger than 36' year olds (n=41, mean 2.93, SD .88) (Sig .012). The age group 26-35 (n=16, mean 2.81, SD .98) shows less interest than the other groups.

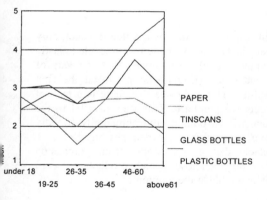

Fig 7 Recycled materials/age
1= Not at all, 5= As much as possible
Paper (n=85), Tins & cans (n=81), Glass
Bottles (n=85), Plastic bottles (n=79)

Question 7 - Do you recycle any of the following material; paper, tins/cans, glass bottles and plastic bottles?

Fig 7 shows that the age group 46-60 (n=21) seems keener on recycling compared to other age groups. The data indicates that there is no significant difference between female and male for recycling. The results indicate that the age group 26-35 (n=17) does recycle less, people aged 46-60 (n=21) do recycle most and the 36-45 (n=15) group seem ambivalent in being both least and most keen.

9

SUMMARY OF THE FINDINGS

The survey was conducted to find out about consumer reactions towards recycled materials for clothing, their shopping behaviour and their environmental awareness, especially in relation to recycling. It also aimed to identify the potential market for clothes made from recycled materials. The outcome is beneficial to the designers, buyers and manufacturers who might consider using recycled material for their products. However, the survey has been relatively small in size due to limitations of time and resources. Nonetheless some key issues have arisen.

General reactions

75.6% of the respondents have favourable reactions towards clothes made from recycled materials. Within this, 51.1% show that they are strongly positive towards recycled materials for clothing. The age group 26 to 45 are more positive than the rest of the age groups.

Pricing

It seems people are not prepared to purchase recycled products if the price of products exceeds more than 10% compared to that of products made from new materials. This suggests that recycled clothes should be priced the same as clothes made from new materials. This result confirms the findings of Buwalda (2001)[25], Rowledge et al. (1999)[26] and Mackenzie (1997)[27]. They noted that consumers are not willing to pay more just because of the product's environmental benefit. Environmental benefit should be extra value along with a high standard of performance, quality and value for money.

Awareness of environmental clothes

Considering the high rate of 83.2 per cent of respondents that own a fleece jacket, only 42.6 per cent of respondents knew that a fleece jacket can be made from recycled plastic bottles and 10.1 per cent of respondents knew a clothing company that is aware of environmental issues. Nobody mentioned Patagonia (USA) who produced the first fleece jacket made from recycled plastic bottles. This indicates that there is a lack of information / awareness towards the products we purchase. This runs parallel to the findings that only 3.4 per cent of respondents have bought environmentally friendly clothes in the past and 9.1% consider the environmental impact when they buy clothes.

Thus it seems that the biggest obstacle that averts people from purchasing environmentally friendly clothes is that they do not know if such products exist. In addition, eco-clothes are not easily or widely available at the moment.

Clothing purchase decisions

The key drivers for a clothing purchase decision, especially knitwear, are 'colour' and 'design'. 'Touch' is also important, particularly for women. It seems that the age group 36 plus care more about 'quality' and that consumers often associated 'quality' with 'branded' clothes. In order to investigate these shopping behaviours further, more research into the particular area of spending would be necessary as spending may vary according to socio-economic status.

Awareness of environmental issues especially recycling

60.5 per cent of the respondents have bought products marketed as environmentally friendly in the past. The data suggested that the age group 36 plus are keen on informing themselves about environmental issues. The age group 46–60 are the most keen on informing themselves as well as doing recycling.

A potential market

According to Marquardt[28], women in their 30-50's in U.S. and Europe are keen on products which have environmental benefits. This survey's results reflect Marquardt's findings. The age group 26 to 45 has a more positive reaction towards clothes made from recycled materials than the other age groups.

Nonetheless, we should not conclude hastily. The survey results show that particularly the age group 26-35 seem least keen on environmental issues, despite their positive reaction towards recycled materials. In comparison while the age group 46-60 do inform themselves about environmental issues and do more recycling than the other age groups, have a less favourable attitude towards recycled materials for clothes than the age group 26-45. They may associate recycled materials with lower quality, since this age group seeks 'Quality' for their clothes more than any other age group.

Overall, the survey result suggests that a major campaign to raise awareness towards 'recycled materials' may be necessary in order to inform the public of the potential of recycled materials.

Limitation of the survey

The data in the survey were collected in the city centre of Newcastle-upon-Tyne and it may be necessary to take into consideration that Newcastle has a high percentage of students in its population. During the process of the data analysis, it was necessary to sum up the data due to the small sample size in particular the age groups. Therefore, some of the results have been explained through two or three age groups rather than the six age groups that were initially used in the questionnaire. Further research would be required if there was a need to acquire more detailed outcomes such as socio–economic status in relation to detailed age groups.

CONCLUSIONS

This study was conducted in order to identify possibilities for clothes with recycled content by exploring public perceptions towards recycled materials and their awareness of environmental issues. The findings that have been derived from the study could be useful indicators for clothing industries that may be using recycled materials for their products.

There could be a few scenarios that would encourage wider use of recycled materials in the clothing industry; such as enforcement by legislation, low prices of recycled materials and an increased demand from consumers. Obviously it would be an ideal situation if these three conditions were to occur simultaneously. Enforcement by legislation and low prices of recycled materials are objectives that could be achieved in a relatively short time. However, these favourable conditions once achieved could as easily be overturned by changes of government policy or a sudden change in oil prices as has occurred during the 2004-2006 period. In both scenarios there is little space for the consumers to control the situation.

On the other hand, creating demand from consumers for recycled materials would need time and effort. People need to realise that we cannot carry on consuming huge amounts of resources and not recycle. However, the rapid growth of certified fair trade goods in the UK indicates a change of consumer shopping behaviours. People have started to show an interest in the origin of products and in their production history. Thirty years ago Schumacher noted that 'The buyer is essentially a bargain hunter' and his main concern is to obtain the highest valued goods for the slightest spending, rather than where or how the goods were produced.[28] Nowadays we are still bargain hunters, but there are signs that we want to take more responsibility for our shopping. Vidal remarks that the reason for this trend is the availability of information. People inform themselves and the simple shock of knowledge can inspire action.[29]

If the availability of information is the key for a change in consumer attitudes, recycled materials may have potential. The survey revealed that people do not know if such products exist, especially for clothing. For instance the age group 46-60 may become more sympathetic towards recycled materials for clothing as they already appear to be keen on informing themselves about environmental benefit. The reason they are not as positive towards clothes made from recycled materials as the age group 25-46 may stem from an association of recycled materials with low quality. If people started realising that recycled materials can be as good quality as virgin materials, there might be an increase in demand for such products and with it an increase in choice at a reduced price.

Consumers need clearer information about products that have environmental benefits and these products would also need to be more readily available. However availability of recycled clothes and assurance of their quality alone would not suffice to convince the consumer. The survey outcomes clearly indicate that people choose clothes by 'colour' and 'design' initially. If 'colour' and 'design' do not appeal to the consumers, there is little chance that they would purchase environmentally friendly clothing. It is thus important that products from recycled materials should be treated as being equal to products produced from virgin materials by designers, buyers and manufacturers in order to guarantee the same standards as the consumer demands from normal clothes.

This research will continue in the future to determine the designers, buyers and manufacturers' reaction towards recycled materials in the clothing industry. In order to have products, which contain recycled materials available for consumption, it is vital to promote awareness of the key players in the industry who are working in this area. This ought to improve the reaction of the industry towards recycled materials and sould eventually lead to the creation and opening of new markets for recycled materials in the textile and clothing industry.

REFERENCES

1   DEFRA (Department for Environment, Food and Rural Affairs). Waste and Recycling Bulletin. *Municipal Waste Management Statistics 2001,* 22 May 2003. http://www.defra.gov.uk/environment/statistics/des/waste/index.htm

2   Ibid

3   DEFRA (Department for Environment, Food & Rural Affairs) Limiting Landfill: A Consultation paper on limiting landfill to meet the EC Landfill Directive's targets for the landfill of biodegradable municipal waste.
http://www.defra.gov.uk/environment/waste/strategy/landfill/2.html#1
(accessed 2001 July 5).

4   R J Ehrig, ed. *Plastic Recycling Products and Processes,* Hansen Publishers, 4, 1992

5   British Plastics Federation
http://www.bpf.co.uk/bpfissues/Waste_Management.cfm (accessed on 18/05/04)

6   Scott Wilson and Save Waste and Prosper Ltd. (SWAP). *Plastic Bottle Recycling in the UK.* The Waste and Resources Action Programme (WRAP) March 2002, p3
http://www.wrap.org.uk/sell_market_studies.asp. (accessed on 05/08/03)

7   Ibid.

8   Ibid.

9   Sutherland, D. Gregg. 'Market Development: Problems and Solutions', *RECYCLING HANDBOOK,* 2nd edition. McGraw-Hill. 2001. 7 p1.

10   Ibid p8.

11   RECOUP. Survey 2000 Report. October 2000
http://www.wasteonline.org.uk/resources/Materials/RECOUP_Survey2000Report.htm
(accessed on 13/08/03)

12   C Cundy, *'Poor quality' perception hampers plastics recyclers'.* Plastic & Rubber Weekly. 4 July 2003. p4
www.wrap.org.uk/publications/
StandardsAndSpecificationsAffectingPlasticsRecyclingInTheUK.pdf
(accessed on 13/08/03)

13   Y Nakano, 'Eco-Clothing:The potential market application for recycled plastic materials as fashion clothing', Cooper, Rachel and Branco, Vasco ed. *d3 desire designum design.* 4th European Academy of Design. Conference proceedings. Universidade de Aveiro. 2001.

14   D Myers and S Stolton, *Organic Cotton,* Guildford: Intermediate Technology. 1999, 101-102.

15 Y Nakano, 'Eco-Clothing: The potential market application for recycled plastic materials as fashion clothing', Cooper, Rachel and Branco,Vasco ed. *d3 desire designum design.* 4th European Academy of Design. Conference proceedings. Universidade de Aveiro. 2001. p.519

16 D Myers and S Stolton, *Organic Cotton,* Guildford: Intermediate Technology. 1999. p103.

17 A Kierstan, Organic baby clothing company plans growth. Drapers record. 02/14/2004. p9.

18 E Musgrave, Hamnett to return with eco-friendly fashion. Drapers record. 01/24/2004.

19 D Myers and S Stolton, *Organic Cotton,* Guildford: Intermediate Technology. 1999. p10.

20 B McCormack and E Hill, *Conducting a Survey, The SPSS Workbook.* International Thomson Business Press, 1997, p30.

21 Census 2001. http://www.newcastle.gov.uk/pr.nsf/a/census2001home

22 B McCormack and E Hill, *Conducting a Survey, The SPSS Workbook.* International Thomson Business Press, 1997, p72.

23 Ibid. p125.

24 W Gordon, 'Goodthinking', A Guide to Qualitative Research. Admap. 1999. p35.

25 T Buwalda, 'Plastics'. Chapter 14. *RECYCLING HANDBOOK.* 2nd edition. McGraw-Hill. 2001. P9.

26 Rowledge, R. Lorinda, Russell S.Barton, Kevin S. Brady. Case 4-Patagonia. *Mapping the Journey: Case Studies in Strategy and Action towards Sustainable Development.* UK: Greenleaf. 1999.p98.

27 D Mackenzie, *GREEN DESIGN – Design for the Environment.* 2nd edition. Laurence King, 1997, p23.

29 E F Schumacher, *SMALL IS BEAUTIFUL: A Study of Economics as People Mattered.* ABACUS. 1973.p36.

30 J Vidal, *Retail therapy - Awareness of how and where goods are produced has soared - and so has the fair trade movement.* 2003. http://society.guardian.co.uk/societyguardian/story/0,7843,902614,00.html (accessed on 12/06/03)

ACOUSTIC AND MECHANICAL PROPERTIES OF UNDERLAY MANUFACTURED FROM RECYCLED CARPET WASTE

[1]Ian Rushforth, [1]Kirill Horoshenkov, [1]Siow N. Ting and [2]Mohsen Miraftab

[1]School of Engineering, University of Bradford, Richmond Road, Bradford, BD7 1DP, U.K.
[2]Centre for Materials Research & Innovation, The University of Bolton, Deane Road, Bolton, BL3 5AB, U.K.

ABSTRACT

In the U.K., carpet waste from manufacturing and fitting operations is estimated to be around 10.5 million m^2 or £70 million per year (Miraftab *et al* 1999). The majority of this waste is either land-filled or incinerated. Growing public concern for the environment and increased landfill taxation is forcing manufacturers to look into alternative uses for their waste. An earlier paper (Rushforth *et al* 2003) demonstrated that it is possible to use recycled carpet waste to manufacture underlay with impact sound insulation performance that is comparable to that of commercially available underlays. The current paper summarises this earlier work and describes the results of standard testing of the optimised recycled underlay and commercial products for impact sound insulation performance (ISO 140-8) and for other standard textile properties (in accordance with BS 5808). Industrial-scale manufacturing trials for the recycled underlay are also described. The results of static loading ('Instron') testing indicate that the use of an appropriate backing material (scrim) in the manufacturing process is likely to improve further the impact performance of the recycled underlay, by spreading the applied load laterally across the material. It is also demonstrated that using a scrim improves the tensile strength of the underlay.

INTRODUCTION

Manufacturers are increasingly looking into alternative uses for their waste output as landfill tax rates increase and public concern for the environment grows. Recent studies by Vitamvasova *et al* [1] and Swift & Horoshenkov [2, 3] have shown that polymeric granulates and fibres, as found in industrial and post-consumer material waste handling processes, may be recycled into materials that have desirable acoustic and physical properties. These novel materials can provide alternatives to virgin products in a number of commercial and environmental noise control applications, including building, automotive, business services and traffic noise abatement. The use of recycled materials reduces the manufacturing costs, the demand for raw materials and the required energy.

In the U.K., carpet waste from manufacturing and fitting operations is estimated to be around 10.5 million m^2 or £70 million per year [4], the majority of which is land-filled and the remainder is incinerated. Miraftab *et al* [4, 5] devised a process that could form carpet waste into a viscoelastic material with potential applications as a carpet underlay. Preliminary tests conducted on samples of these underlays [6] indicated that the materials could improve impact sound insulation in flooring applications. This would be beneficial for noise control in buildings in accordance with noise legislation such as the UK Building Regulations Approved Document E [7]. Such an approach

15

also addresses the problem of disposal of carpet waste, thus reducing environmental pollution and usage of virgin material, as well as potentially reducing production costs.

This paper outlines the results of research [8] into the impact sound insulation performance of recycled underlays that were laboratory produced from granulated carpet tiles. A comparison with commercially available acoustic underlays was carried out. Industrial-scale manufacturing trials also took place. An optimised recycled acoustic underlay product was developed and submitted for standard acoustic testing at accredited facilities. The static loading behaviour of the samples was also investigated, and the results enabled a better understanding of the impact results and highlighted the potential to improve further the acoustic performance of the recycled underlay.

MANUFACTURING PROCESS

Laboratory-scale samples

The manufacturing of underlay samples in the laboratory from recycled carpet waste is outlined below. For a detailed description of the manufacturing methodology, see [8].

PVC-backed carpet tiles with nylon/polypropylene pile were passed through a granulator and cyclone separation system, yielding granular (backing) and fibrous (pile) components of carpet waste. The separated fibrous and granular components were then mixed together again in controlled ratios and consolidated using a pre-foamed styrene-butadiene rubber (SBR) binder. The mixture was then spread into a mould, ensuring a uniform material thickness, and the mould was placed in an oven at 130°C for approximately 1½ hours until the sample had dried and cured.

A number of samples were manufactured in the laboratory according to the above method, to a standard thickness of 10mm, whilst varying parameters such as grain:fibre ratio of the dry component, binder concentration, particle size distribution, etc.

Industrial-scale production

The laboratory sample that performed best in indicative impact testing, U2, [8] was reproduced on an industrial scale at a textile mill of *Anglo Felt Industries Ltd*, in order to investigate the viability of industrial-scale production of underlays from recycled carpet waste. The optimised formulation was coated onto a polypropylene carrier material (scrim), which was conveyed through a fan-assisted industrial heater. The coating assembly could be adjusted to control the thickness of the material prior to heat treatment, and the product emerged from the oven having dried and cured at a considerably greater rate that that observed under laboratory conditions.

The industrial trials indicated that production of underlays of a homogeneous nature and constant thickness, from recycled carpet waste, is potentially viable on an industrial scale. However, the thickness of the resultant materials from these initial trials was in the range of 5-6mm, rather than the standard thickness of 10mm.

EXPERIMENTAL METHODOLOGY

Impact rig

Underlays designed for use in acoustic flooring systems must comply with Building Regulations Approved Document E [7] and ISO 140 part 8 [9]. However, it is both impractical and expensive to conduct these tests at accredited facilities for a large number of samples. In order to enable a comparative assessment of impact sound insulation performance, a small test rig (see **Figure 1**) was constructed which allowed for much smaller specimens to be tested efficiently in the laboratory.

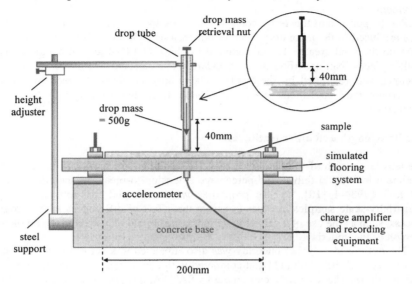

Figure 1: *Schematic diagram of the impact transmission test rig.*

The experimental methodology is summarised below; for a full description see [8].

The recycled underlay samples were fixed to a timber plank, designed to simulate a typical flooring system. The samples were then subjected to impacts of constant force from a brass cylinder of mass 500g, dropped from a height of 40mm in the tube as shown in **Figure 1**. An accelerometer attached to the underside of the plank measured the acceleration level of transmitted vibration through the structure.

Impact events were recorded digitally and subjected to Fourier analysis to calculate octave band spectra of the relative acceleration levels. This procedure was carried out for each laboratory-produced carpet waste underlay sample, as well as for several high-performing acoustic underlays available commercially. By comparing the octave band spectra attained, the relative degree of attenuation of impact sound by each sample was assessed. The lower the magnitude of the transmitted vibration, the better performing the sample under test. Accordingly, the acoustic performance of the recycled carpet underlays was determined.

Standard test for impact sound insulation

The standard test for impact sound reduction of floor coverings, ISO 140 part 8 [9], was carried out on the optimised recycled underlay U2 and on two of the commercial underlays that had performed well in indicative impact tests. This was conducted at specialist testing facilities at Salford University. The test procedure is as follows.

Each underlay is installed on a concrete floor above a reverberant chamber. A tapping machine containing five impact hammers is placed on the underlay, and used to generate impact sound. The tapping machine is also operated when placed on the bare floor. Sound pressure levels within the reverberation room below are measured and spatially averaged using an array of fixed microphones. A correction is made for the reverberation time of the receiver room. The result is given in terms of the weighted reduction of impact sound pressure level, ΔL_w, averaged over three 1.0m x 0.5m specimens.

For the purpose of this test, twelve 10mm-thick specimens of optimised underlay U2 were produced in the laboratory, each of dimensions 375mm x 330mm, i.e. sufficient to make up the total area of 1.5 m^2 required by the ISO 140-8 test, although in separate smaller pieces than specified in the standard. The selected 375mm x 330mm size of specimen was restricted by the available laboratory facilities at Bradford University. The commercial specimens were also 10mm thick but were cut to the specified dimensions.

Tensile strength and other textile tests

The tensile strength of the optimised recycled underlay was measured both in terms of maximum loading to failure and percentage extension-to-break, in accordance with BS EN ISO 13934-1 [10]. These properties were measured for both laboratory and industrially-scale samples in order to assess the effect of differences in the production process, such as the use of a carrier material (scrim) in the industrial trials, on the tensile strength of the underlay.

The optimised recycled underlay was also subjected to a series of standard tests in accordance with BS 5808 [11] and compared with commercially available underlays of similar calibre. The tests were carried out by SATRA, a specialist testing house for floor coverings.

Static loading

In order to understand more fully the behaviour of the samples in impact tests, the static loading behaviour of the test specimens was investigated using an 'Instron' mechanical loading machine. Stress versus strain curves were plotted during compression of the samples up to a high value of stress of around 800 kPa (approximately equivalent to a load of 80 kg transmitted through an area of 10^{-3} m^2 i.e. an average adult stepping on the floor with the heel of a shoe).

RESULTS

Impact rig

A detailed description of the results of laboratory impact testing and optimisation of the recycled carpet underlay is given in [8]. **Figure 2** shows the most salient results, with the performance of the optimised recycled sample U2 plotted against two of the best-performing commercial underlays. Note that all samples were 10mm thick, and that low amplitude of transmitted vibration corresponds to effective impact sound insulation performance.

Overall, the impact results from laboratory tests suggest that the impact sound reduction capability of the optimised recycled underlay U2 is similar to, and in some cases considerably better than, the commercial products.

18

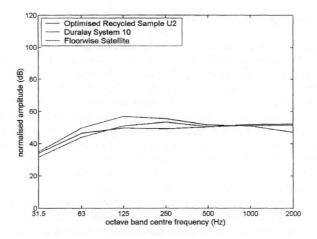

Figure 2: *A comparison of the impact sound insulation performance of the optimised recycled underlay U2 and two commercially available underlays (middle line shows optimised recycled underlay U2).*

Impact testing of one of the industrially produced samples from the *Anglo Felt* trials was also carried out. Two 5mm-thick specimens were stacked up to obtain a thickness of 10mm, in order to allow a controlled comparison with the 10mm-thick laboratory produced sample of the same formulation. The results shown in **Figure 3** demonstrate that the transmitted vibration spectra for this industrial sample and for the laboratory specimen of identical formulation were very similar. This suggests that it should be possible to reproduce the acoustic performance of the laboratory samples on an industrial scale.

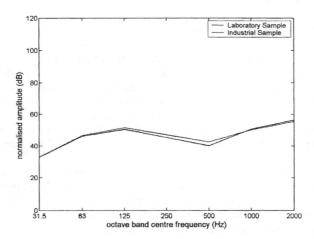

Figure 3: *A comparison of the impact sound insulation performance of industrial-scale and laboratory-scale samples of recycled underlay (top line shows industrial sample).*

Standard test for impact sound insulation

The value of ΔL_w for the optimised recycled underlay U2, obtained in accordance with ISO 140-8, was found to be 28 dB; this compares favourably with the minimum value of 17 dB stipulated in the Building Regulations [7]. In addition, the performance of U2 showed no sign of deterioration with time over the duration of the test.

Two commercial acoustic underlay samples were also tested on the same occasion. One of these, Duralay System 10 (which had performed similarly or slightly worse than U2 on the indicative impact rig in the laboratory – see **Figure 2**), yielded a slightly higher value of $\Delta L_w = 31$ dB. The other commercial sample, Floorwise Satellite (which had performed better overall in indicative tests) yielded a noticeably better value of $\Delta L_w = 37$ dB in ISO 140-8 testing. Both commercial samples also showed no deterioration in performance over the duration of the test.

Thus the optimised recycled underlay U2 performed reasonably well in this standard test, but not as well as the two commercial samples tested. There are, however, two factors that should be taken into consideration, which may have reduced the performance of U2. Firstly, the sample U2, unlike the commercial samples, was not tested in the form of three whole pieces (each 1.0m x 0.5m) as recommended in the ISO standard. Whilst care was taken to avoid any of the tapping machine's hammers falling on a 'seam' during the test, the effect of the discontinuous specimens may have been to degrade performance by several decibels.

Secondly, both commercial underlays have stiff backing materials (scrim), whereas the recycled underlay U2 produced in the laboratory had a thin, highly permeable, low strength backing. The Floorwise Satellite commercial underlay in particular has a crepe paper backing. The effect of this stiff backing may be to spread the load over a wider area than the immediate point of impact during testing, resulting in the highest value of ΔL_w in the ISO test. This 'load-spreading' effect would not be observed in the case of sample U2, due to the very low stiffness of the scrim on which the wet formulation was spread during sample production.

Tensile strength and other textile tests

The tensile properties of laboratory and industrially-produced samples of the recycled underlay were determined. As shown in **Table 1**, the tensile strength of the industrially-produced sample is considerably greater than that of the laboratory sample. It would appear that the carrier fabric (scrim) used during the industrial trials contributes considerably to the overall tensile strength of the underlay, both in terms of maximum loading to failure and percentage extension to break. This is confirmed by the results shown in **Table 2**, which shows the tensile properties of the backing material alone, in both machine and cross machine direction.

The optimised recycled underlay and commercial underlays were also tested in accordance with BS 5808 [11]. The results of these tests are outlined in full elsewhere [12], but to summarise the optimised recycled underlay was rated as L/U - Luxury use, domestic/contract, i.e. for use where high energy absorption is desirable. In contrast, the commercial underlay that had performed best in the ISO 140-8 test for impact sound insulation, namely Floorwise Satellite, was rated as GD/U - General Domestic use.

Industrial Sample						Laboratory Sample		
Mean Load at Maximum Load (kN) (Machine Direction)	S.D.	Coef. V.	Mean Load at Maximum Load (kN) (Cross Machine Direction)	S.D.	Coef. V.	Mean Load at Maximum Load (kN)	S.D.	Coef. V.
0.511	.0157	3.08	0.38	.0593	15.27	0.0237	0.001	4.34
Mean Displacement at Maximum Load (mm)	S.D.	Coef. V.	Mean Displacement at Maximum Load (mm)	S.D.	Coef. V.	Mean Displacement at Maximum Load (mm)	S.D.	Coef. V.
64.78	7.97	12.31	31.16	3.29	10.54	20.18	1.12	5.56

Table 1: *A comparison of the tensile properties of industrial-scale and laboratory-scale samples of recycled underlay.*

Polypropylene Scrim					
Mean Load at Maximum Load (kN) (Machine Direction)	S.D.	Coef. V.	Mean Load at Maximum Load (kN) (Cross Machine Direction)	S.D.	Coef. V.
0.603	.6027	2.27	0.4114	0.0250	6.09
Mean Displacement at Maximum Load (mm)	S.D.	Coef. V.	Mean Displacement at Maximum Load (mm)	S.D.	Coef. V.
65.92	2.38	3.61	28.20	2.70	9.57

Table 2: *Tensile properties of the polypropylene scrim used as a backing material during industrial trials.*

Static loading

The 'load-spreading' effect described earlier was investigated in the laboratory at Bradford University, by studying the static loading behaviour of the test specimens using the 'Instron' mechanical loading machine. **Figure 4** shows stress versus strain curves during compression of the samples up to 800 kPa.

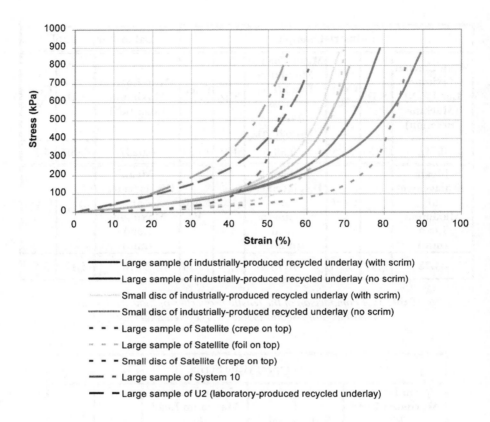

Figure 4: *Stress versus strain curves, illustrating the static loading behaviour of the optimised laboratory and industrially produced recycled underlays along with two commercial underlays, and the influence of a stiff backing on the results.*

The shorter dashed curve at the far right of the figure shows the behaviour under compression of a large sample[1] of the Satellite commercial underlay placed the right way up (i.e. crepe paper backing on top), whilst the lighter dot-dash curve at the left of the figure applies to a large sample of System 10 commercial underlay and the dark longer dashed curve at the left of the figure applies to lab-produced sample U2 (large sample).

It can be seen that for a high value of stress (e.g. 800 kPa), the Satellite sample can take up a much higher degree of strain than the other two samples, which explains its superior performance in impact sound insulation tests. The System 10 and U2 samples display similar stress-strain curves, suggesting similar impact reduction abilities. (The sample U2 has undergone slightly higher strain for a given stress than System 10; U2 performed slightly better in small-scale impact tests but System 10 yielded better performance in the ISO 140-8 tests.)

---

[1] A large sample may be defined as one that extends radially beyond the circumference of the compression 'footprint' of the Instron machine by a significant distance i.e. at least several times the radius of the 'footprint'.

However, **Figure 4** also shows (short dashed curve towards the left of the figure) that placing a large sample of Satellite with the crepe paper backing underneath reduces significantly the degree of strain at a given high stress level. The thin foil backing exposed on the top is more malleable than crepe paper and less stiff, leading to a reduced capability to spread the load over a wider area.

This is demonstrated further by the short dashed curve in the middle of the figure (**Figure 4**), which shows the behaviour under compression of a small disc-shaped sample of Satellite (crepe paper backing on top), which was cut so that the diameter of the sample was equal to that of the Instron 'footprint', i.e. the area of contact on the sample during the compression test. In this case, there was no possibility for the load to spread over a wider area rather than the immediate point of contact. Consequently, the compressional modulus behaviour was similar to that of the other two samples (U2, long dashed curve at the left, and System 10, dot-dash curve at the left).

Thus it may be inferred that the compressional modulus values of the core material of the investigated samples are similar. Furthermore, as reported elsewhere [12], the recycled underlay has a higher value for loss factor (tan δ) than any of the commercial underlays, possibly due to the interaction between the grains and fibres. A combination of low compressional modulus and high loss factor is indicative of good impact sound insulation performance.

Further Instron measurements were carried out on the recycled underlay produced in industrial trials at *Anglo Felt*, which was backed with a superior, stiffer scrim than the laboratory sample U2. Note that the thickness of the industrially produced specimens was in the range of 5-6mm, rather than the standard thickness of 10mm. Therefore, two 5mm thick specimens were selected for each run of the Instron and were stacked up to obtain the required 10mm thickness. It is possible that in this arrangement, some of the strain may be dissipated by air gaps between the layers or by slippage across the lamination (between the doubled-up pieces).

Both large and small circular samples (for definition of 'large' and 'small' see earlier footnote) were tested with and without the scrim on the upper surface during the tests. The solid curves (first and second from the right respectively) in **Figure 4** correspond to the larger samples with and without the scrim, respectively. The presence of the scrim increases the strain dissipated by the sample for a given applied stress. This suggests that spreading of the applied load across a wider area than the immediate point of contact is taking place as a result of the stiff backing material.

Furthermore, comparison of the solid and the shorter dashed curves at the right suggests that the industrially produced recycled underlay sample with scrim, is capable of taking up similar amounts of strain at high stress levels as the commercial underlay Satellite (correctly installed with crepe backing on top). It may be inferred that the recycled underlay with stiffer scrim produced on an industrial scale could yield improved performance in the ISO 140-8 test compared with the lab-scale recycled underlay, and perhaps comparable with the high performance standard of Satellite in the ISO test ($\Delta L_w = 37dB$).

The fourth and third solid curves from the right in **Figure 4** correspond to the mechanical behaviour of the small samples of the industrially produced recycled underlay, with and without scrim respectively. The behaviour of these two samples is similar. Additionally, the two large samples are capable of dissipating a greater strain per applied stress than either of the small samples. These results confirm that the lateral load-spreading effect occurs rather efficiently in the large sample with scrim, but also suggest that a lesser degree of lateral distribution of load can take place in the large sample without scrim. When small discs of underlay are subjected to loading, there is no opportunity for the stress to be distributed laterally whether a scrim is present or not.

In conclusion, the use of a high quality, stiff backing when producing the recycled underlay, as well as facilitating the manufacturing process and improving the tensile strength and structural integrity of the material, would serve to enhance the impact sound insulation performance of the product by increasing the degree of lateral 'load-spreading' occurring within the underlay.

CONCLUSIONS

It has been demonstrated that it is possible to recycle carpet waste into an acoustic underlay product. Underlay samples were developed by granulating PVC-backed, nylon/polypropylene-pile carpet tiles and binding the waste together with an SBR binder. The construction of a specially designed rig allowed the impact sound insulation capability of the recycled materials to be assessed and compared with commercially available underlay products.

The recycled underlays manufactured in this way performed well in the ISO 140-8 test for impact sound insulation of floor coverings, and in the BS 5808 standard textile tests for underlays. Although some commercial underlays performed better in the ISO 140-8 standard test, the use of an appropriate backing material (scrim) in the recycled underlay should improve the impact performance further, by spreading the applied load laterally across the material, as well as increasing the tensile strength of the underlay.

Initial industrial trials demonstrated that production of homogeneous recycled carpet underlays of consistent thickness is technically viable on an industrial scale, and that it should be possible to reproduce the acoustic performance of the laboratory samples at that scale.

Landfill tax rates are on the increase but by adopting the process outlined in this paper, carpet manufacturers could reduce their landfill costs by recycling their waste output. Therefore, since the raw material is cheap and readily available, and the liquid binder used (SBR) is relatively inexpensive, the production of recycled acoustic underlay could be commercially viable.

To summarise, this study demonstrates that recycling carpet waste to produce quality acoustic underlays with desirable impact sound insulation characteristics is technically feasible and a viable alternative to landfill or incineration.

ACKNOWLEDGEMENTS

This research was sponsored by the UK Department of Trade and Industry under its Waste & Resources Action Programme (WRAP). The authors also wish to acknowledge the input of Dr Mark Swift, who worked on the initial stages of the project and the technician, Mr Clive Leeming, who helped to design and built the impact transmission rig and who provided essential technical support.

REFERENCES

1. E. Vitamvasova, S Vasult, D Gerza, and P Bris, 'A noise and vibration protection by mixed polymer waste', *Proceedings of Internoise '96, University of Liverpool,* 1996 1835-9.

2. M Swift, P Bris, and K Horoshenkov, 'Acoustic absorption in re-cycled rubber granulates', *Applied Acoustics*, 1999 **57** 203-212.

3. M Swift and K Horoshenkov, 'Acoustic and physical properties of recycled granular foams', *CD-ROM Proceedings of the Euronoise 2001 Conference, Patras, Greece* 2001.

4. M Miraftab, R Horrocks and C Woods, 'Carpet waste, an expensive luxury we must do without', *AUTEX Research J*, 1999 **1** 1-7.

5. A Taylor, 'Novel underlays from carpet waste', Ph.D. thesis, The University of Bolton, U.K in progress.

6. M Swift, 'Acoustic optimisation of newly developed underlay system', Acoutechs Ltd / Bradford University. Client Report CR3/MJS/02/02 2002.

7. Department of Environment, 'Resistance to the passage of sound', The Building Regulations 2000 Approved Document E 2003 edition.

8. I Rushforth, K Horoshenkov, M Miraftab and M Swift, 'Impact sound reduction and viscoelastic properties of underlay manufactured from recycled carpet waste', submitted for publication in *Applied Acoustics* 2003.

9. British Standards Institute. 'Acoustics - measurement of sound insulation in buildings and of building elements – Part 8: Laboratory measurements of the reduction of transmitted impact noise by floor coverings on a heavyweight standard floor', BS EN ISO 140-8 1998.

10.   British Standards Institute, 'Textiles - Tensile properties of fabrics – Part 1: Determination of maximum force and elongation at maximum force using the strip method', BS EN ISO 13934-1 1999.

11. British Standards Institute, 'Specification for underlays for textile floor coverings', BS 5808 1991.

12. WRAP report, 'Recycling carpet waste into acoustic underlay for commercial production', Waste & Resources Action Programme, available at www.wrap.org.uk 2004.

CARPET FIBER RECYCLING TECHNOLOGIES

Youjiang Wang
School of Polymer, Textile & Fiber Engineering, Georgia Institute of Technology
Atlanta, Georgia 30332-0295 U.S.A.

ABSTRACT

Significant progress has been made in waste minimization and pollution prevention in textile manufacturing processes. Because most carpets and textiles are for replacement, recycling post-consumer fibrous waste should be an integral part of sustainability for textile products. Currently in the United States alone, over 2 million tons of post-consumer carpet waste is discarded into landfills each year, and the amount is expected to increase to over 3 million tons by 2012. Very little post-consumer carpet at present is recycled. To establish a sustainable commercial network to recycle fibrous waste, operations based on different technologies must coexist such that different types of materials collected can be recycled to the greatest extent. This paper reviews technologies for carpet waste recycling.

INTRODUCTION

The US carpet industry is responsible for 45% of world's carpet production [1]. The great majority of carpet is made by tufting. In this typical construction, face fibers are tufted on to a primary backing and locked into place by an adhesive layer (latex filled with $CaCO_3$) and a secondary backing fabric. The backing fabrics are normally made of polypropylene, while the face fiber is often made of nylon, although polyester, polypropylene and wool fibers are also used for some carpet.

The entire US textile industry consumes 8 million tons of fibers with 1.6 million tons used in carpet manufacturing. Most of the fibrous products are purchased as replacement, resulting in over 7 million tons of fibrous waste (apparel, carpet, and textiles) being disposed of in landfills each year. Due to demand for recycled products, landfill costs and environmental regulations, treating this large amount of fibrous waste as a resource for raw material by converting it into useful products is the most desirable solution. The industry has taken a proactive approach by partnering with private and public sectors to divert carpet waste from landfills. The Carpet America Recovery Effort (CARE) has set a goal of recycling 0.7 million tons of carpet waste (22%) per year by 2012 in the U.S. There has been significant effort on developing technologies to recover materials from carpet waste, including some large industrial operations to convert carpet waste into plastics and chemicals. Unfortunately some of the operations discontinued for economic reasons. To have a sustainable commercial network to recycle fibrous waste, operations based on different technologies must coexist and work in a concerted manner. This paper reviews technologies for carpet waste recycling [2]. The basic approaches are chemical (e.g., depolymerization), physical (e.g., solvent, supercritical fluid), and mechanical (e.g., size reduction and density differentiation).

COLLECTION AND SORTING

Collecting and sorting are necessary steps for carpet recycling. After collection, the post consumer carpet (PCC) is transported to sorting facilities, where the collected PCC is either manually or automatic sorted according to the face fiber. For many recycling processes such as nylon depolymerization and polymer resin recovery, it is desirable or required to sort the feedstock according to the fiber type. A melt pointing indicator is an inexpensive instrument that can identify most fiber types, but it is generally slow and cannot distinguish between nylon 66 and polyester. Infrared spectroscopy is a must fast and accurate technology. A typical instrument consists of an A/C powered base unit for data acquisition, analysis and display, and a probe connected to the based unit via a fiber-optical cable. Such units are suitable for carpet sorting in a central warehouse. A portable infrared spectrometer has been developed by Kip et al [3], which is a lightweight, battery operated unit. It is designed to identify the common carpet face fibers: nylon 6, nylon 66, polypropylene, polyester, and wool. Unidentifiable fibers, either due to operating conditions or fiber types other than those in the above list, would be shown as "unknown".

MECHANICAL SEPARATION OF CARPET COMPONENTS

Mechanical methods have been utilized to separate carpet components. One or more segregated components then are recycled into products that generally compete with products produced from virgin polymers.

In a process developed by DuPont [4,5], nylon 66 carpet first is passed through dry processes consisting of a series of size reduction and separation steps. This provides a dry mix of 50-70% nylon, 15-25% polypropylene and 15-20% latex, fillers and dirt. Water is added in the second step where the shredded fiber is washed and separated using the density differences between the fillers, nylon and polypropylene. Two product streams are obtained: 98% pure nylon and 98% pure polypropylene. The recycled nylon is compounded with the virgin nylon at a ratio of 1:3 for making automotive parts.

The United Recycling process [6,7] starts with clipping the face fibers on loop carpet to open the loops. The next step is debonding, in which the carpet is bombarded with a combination of air and steam to loosen the calcium carbonate-filled latex backing. The secondary backing then is peeled off mechanically, exposing the primary polypropylene backing. Next mechanical picks pluck the face fibers. It is claimed that the cost of this process is low and that it yields a product stream with 93-95% pure face fibers. Other devices employing water jet [8], dry ice [9] or mechanical actions [10] for size reduction and separation of carpet have also been reported. Many types of equipment are commercially available for processing textile and carpet waste.

SOLVENT EXTRACTION OF NYLON FROM CARPET

Solvent extraction has also been used to separate the high value nylon from carpet waste. The solvents used are aliphatic alcohol [11], methanol [11], alkyl phenols [12], and hydrochloric acid [13]. Typically, the yield of nylon is high and no degradation of the recycled nylon is observed., but the drawbacks of solvent extraction are the chemicals involved, modest temperature and pressures required, and time required. Some solvents, such as hydrochloric acid, are not recyclable due to reaction with the calcium carbonate filler in the carpet waste.

27

Another approach to separate carpet components is to use a supercritical fluid (SCF) method [14,15]. The solubility of a polymer changes with the variation in pressure and temperature of the SCF. After nylon is dissolved in a solution such as formic acid, a supercritical CO_2 as an anti-solvent is added to precipitate the nylon out of the solution. Both the solvent and the anti-solvent can be recycled.

DEPOLYMERIZATION OF NYLON

Recovery of raw chemicals that can be used to produce virgin-quality nylon is a very attractive approach. Chemical recycling of nylon 6 carpet face fibers has been developed into a closed-loop recycling process for waste nylon carpet [16,17,18,19]. The recovered nylon 6 face fibers are sent to a depolymerization reactor and treated with superheated steam in the presence of a catalyst to produce a distillate containing caprolactam. The crude caprolactam is distilled and repolymerized to form nylon 6. The caprolactam obtained is comparable to virgin caprolactam in purity. The repolymerized nylon 6 is converted into yarn and tufted into carpet. The carpets obtained from this process are very similar in physical properties to those obtained from virgin caprolactam.

The "6ix Again" program of the BASF Corp. has been in operation since 1994. Its process involves collection of used nylon 6 carpet, shredding and separation of face fibers, pelletizing face fiber for depolymerization and chemical distillation to obtain a purified caprolactam monomer, and repolymerization of caprolactam into nylon polymer [19].

Evergreen Nylon Recycling LLC, a joint venture between Honeywell International and DSM Chemicals, was in operation from 1999 to 2001. It used a two-stage selective pyrolysis process. The ground nylon scrap is dissolved with high-pressure steam and then continuously hydrolyzed with super-heated steam to form caprolactam. The program has diverted over one hundred thousand tons of post consumer carpet from the landfill to produced virgin-quality caprolactam [17,18].

Depolymerization of nylon 66 to recover adipic acid and hexamethylene diamine (HMDA) has also been demonstrated [4,20,21,22] but has not been implemented in commercial operation.

GLASS FIBER-REINFORCED THERMOPLASTIC COMPOSITES [23,24]

Recycling post-consumer carpet is difficult due to the co-mingling of incompatible polymers and the general level of contamination encountered. If the polymers are cleaned and separated, the recycled polymers can achieve reasonable property levels to qualify for a variety of applications. Unfortunately, separating certain polymeric products, like post-consumer carpet, can be difficult and costly. The focus of this research is to develop economically attractive technology to recycle post-consumer carpet by using sufficient fiber reinforcement that the fibers dominate the properties of the composite. Post-consumer nylon 6, nylon 66 and polypropylene carpet was shredded and converted into pellets using an NGR A-Class Type 55 VSP Repelletizing system. To prepare laminates the pellets were converted into a coarse powder using a Wiley mill. Powder was interleaved with glass mats to achieve a laminate with 30 or 40 wt % glass. Laminates were also made without the pelletizing step, by first debulking the shredded carpet in a hot press, and then compression molding with glass mats.

Properties from these processes are reported in Table 1. The laminates exhibit mechanical properties similar to commercial glass mat-reinforced thermoplastics (GMT). Further improvements are expected through the addition of compatibilizers and appropriate adhesion promoters.

Table 1. Mechanical Properties of Glass Fiber-Reinforced PCC

PCC material, process and % Glass Fiber	Flex. Strength MPa	Flex. Modulus GPa	Drop Impact J (@4mm thick)
PP debulk-comp mold-30%	54	2.2	23
-pellet-comp mold – 30%	68	4.8	20
-pellet-comp mold – 40%	94	6.2	31
N6 debulk-comp mold – 30%	70	2.0	20
-pellet-comp mold – 30 %	135	4.6	19
-pellet-comp mold - 40%	157	4.2	24
N66 debulk-comp mold - 30%	113	3.2	
-pellet-comp mold - 30%	147	5.7	
-pellet-comp mold – 40%	179	8.1	
PP Azdel – 32% (*)	104	4.6	9
PP Azdel – 40% (*)	146	5.5	10

* Commercial products of Azdel, Inc. (www.azdel.com)

CARPET WASTE FIBER FOR CONCRETE AND SOIL REINFORCEMENT

Nylon and polypropylene fibers from carpet can be used for concrete reinforcement. A laboratory study on concrete reinforcement was carried out using carpet waste fibers (Typical length 12 to 25 mm) at fiber volume fractions from 1-2% [25,26]. Four point flexural test and cylinder compressive test were conducted. In the compressive tests, the plain concrete specimens failed in a brittle manner and shattered into pieces. In contrast, all the FRC samples after reaching the peak load could still remain as an integral piece, with fibers holding the concrete matrices tightly together. In the flexural test, it was observed that the plain concrete samples broke into two pieces once the peak load was reached, with very little energy absorption. The FRC specimens, on the other hand, exhibited a pseudo ductile behavior and fibers bridging the beam crack can be seen. Because of the fiber bridging mechanism, the energy absorption during flexural failure was significantly higher than that for plain concrete. A construction project by the Shaw Industries, Inc. demonstrated the feasibility of using large amount of carpet waste for concrete reinforcement in a full scale construction project.

It has been widely reported that the properties (especially the shear strength) of soil can be enhanced by fiber reinforcement, resulting in a more stable soil structure with improved load-bearing capacities and durability. The feasibility of using shredded carpet waste for soil reinforcement was investigated [27]. A series of laboratory strength and deformation tests are performed to evaluate the relative performance of unstabilized and carpet waste fiber stabilized soils. Test results indicate that under certain conditions such as large deformation, fibers are especially effective in enhancing the performance of the soil. Several promising applications have been identified.

SUMMARY

Large amounts of carpet waste landfilled each year has created social, economical and environmental concerns. On the other hand, there is an abundance of mixed polymers that may be harvested as raw material for the fibers, nonwovens and plastics industries. For example, about 0.8 million tons of nylon, one of the most expensive commodity plastics, can be found in discarded carpet alone in the US each year. There has been significant effort on developing technologies to recover materials from carpet waste, including some large industrial operations to convert carpet waste into plastics and chemicals. To have a sustainable commercial network to recycle fibrous waste, operations based on different technologies must coexist and work in a concerted manner. This paper provides an overview of technologies for carpet waste recycling. It also discusses the research conducted at the Georgia Institute of Technology to develop low-cost composites from post consumer carpet.

REFERENCES

1 The Carpet and Rug Institute, www.carpet-rug.com, 2004.

2 Y Wang, Y Zhang, M Polk, S Kumar and J Muzzy, *Chapter 16: Recycling of Carpet and Textile Fibers*, Plastics and the Environment: A Handbook, Edited by A. L. Andrady (John Wiley & Sons, New York, 2003), 697-725, 2003.

3 B J Kip, E A T Peters, J Happel, T Huth-Fehre and F Kowol, Method of identifying post consumer or post industrial waste carpet utilizing a hand-held infrared spectrometer, US Patent Office, Pat No 5 952 660 September 14, 1999.

4 H P Kasserra, *Recycling of Polyamide 66 and 6,* Science and Technology of Polymers and Advanced Materials, Edited by P. N. Prasad *et al.*, Plenum Press, New York, 629-635, 1998.

5 J Herlihy, 'Recycling in the Carpet Industry', *Carpet and Rug Industry*, 17-25, November/December 1997.

6 J A E Hagguist and R M Hume, Carpet reclaimer, U.S. Patent Office, Pat No 5 230 473 July 1993.

7 J H Schut, 'Big Plans for Carpet', *Plastics World*, p.25, December 1995.

8 M A Howe, S H White, S G Locklear, Method and apparatus for reclaiming carpet components, U.S. Patent Office, Pat No 6 182 913, February 6, 2001.

9 F C Bacon, W R Holland, L H Holland, Method and machine for recycling discarded carpets, U. S. Patent Office, Pat No 5 704 104, January 6, 1998.

10 K Hawn, 'An overview of commercial recycling technologies and textile applications for the products', 6th Annual Conference on *Recycling of Polymer, Textile and Carpet Waste*, Dalton, Georgia, April 30-May 1, 2001.

11 M Booij, J A J Hendrix and Y H Frentzen, Process for recycling polyamide-containing carpet waste, European Patent Office, Pat No 759 456, February 1997.

12 Y H Frentzen, M P Thijert and R L Zwart, Process for the recovery of caprolactam from waste containing nylon by extraction with alkyl phenol, World Patent Office, Pat No 97 03 04, 1997.

13 A K Sarian, A A Handerman, S Jones, E A Davis and A Adbye, Recovery of polyamide from composite articles, US Patent Office, Pat No 5 849 804, 1998.

14 M E Sikorski, Recycling of polymeric materials from carpets and other multi-component structures by means of supercritical fluid extraction, US Patent Office, Pat No 5 233 021, August 1993.

15 A T Griffith, Y Park and C B Roberts, 'Separation and recovery of nylon form carpet waste using a supercritical fluid antisolvent technique, *Polymer-Plastics Technology and Engineering*, 1999 **38**(3) 411-432.

16 P Bajaj and N D Sharma, *Reuse of Polymer and Fibre Waste* in Manufactured Fibre Technology (Ed. Gupta, V. B. and Kothari, V. K.), Chapman & Hall, New York, p. 615, 1997.

17 T Brown, 'Infinity Nylon - A never-ending cycle of renewal', Presentation at 6th Annual Conference on *Recycling of Fibrous Textile and Carpet Waste*, Dalton, Georgia, April 30-May 1, 2001.

18 C C Elam, R J Evan and S Czernik, An integrated approach to the recovery of fuels and chemicals from mixed waste carpets through thermocatalytic processing', Preprint papers - *American Chemical Society, Division of Fuel Chemistry*, 1997 **42**(4) 993-997.

19 BASF Corp., "BASF 6ix Again Program", www.nylon6ix.com, 2001.

20 B Miller, U.S. Patent Office, Pat No 2 840 606, 1958.

21 M B Polk, L L LeBoeuf, M Shah, C -Y Won, X Hu, and Y Ding, *Polym.-Plast. Technol. Eng.* 1999 **38**(3) 459.

22 S Bodrero, E Canivenc and F Cansell, Chemical recycling of polyamide 6.6 and polyamide 6 through a two step ami-/ammonolysis process', 4th Annual Conference on *Recycling of Fibrous Textile and Carpet Waste*, Dalton, Georgia, May 17-18, 1999.

23 J Muzzy, Y Wang, M Satcher, B Shaw, A McNamara and J Norton, 'Composites derived from post-consumer nylon 6 carpet", ANTEC 2003, *Annual Technical Conference of the Society of Plastics Engineers*, May 4-8, Nashville, TN, 2003.

24 J Muzzy, Y Wang, C Hagberg, P Patel, K Jin, S Samanta, L Bryson and B Shaw, 'Long fiber reinforced post-consumer carpet', ANTEC 2004, *Annual Technical Conference of the Society of Plastics Engineers*, May 16-20, Chicago, Illinois, 2004.

25 Y Wang, A Zureick, B S Cho, D Scott, 'Properties of fiber reinforced concrete using recycled fibers from carpet industrial waste', *J of Materials Science,* 1994 **29**(16).

26 Y Wang, 'Utilization of recycled carpet waste fibers for reinforcement of concrete and soil', *J of Polymer-Plastics Technology & Engineering,* 1999 **38**(3) 533-546.

27 J Murray, J D Frost and Y Wang, 'The behavior of sandy soil reinforced with discontinuous fiber inclusions', *Transportation Research Record,* 2000, No. 1714, 9-17.

USE OF WASTE AS RAW MATERIALS: EFFICIENT RECYCLING TECHNIQUES

Simon K. Macaulay

Anglo Felt Industries Ltd., Tong Lane, Whitworth, Rochdale
OL12 8BG, UK

INTRODUCTION

Most of us are generally aware of waste. We now are required to recycle our newspapers and plastic containers, and go to the bottle bank. Some of us take our old clothes to the charity shop.

This is entirely admirable. But what use is put to that material in the recycling process – how does the market operate, especially with regards to textiles. High value materials (e.g. Nomex, Kevlar etc.) have always had a ready recycling route and market outlet but this paper focuses on the economics of low value waste and the way that a company, like Anglo Felt Industries, is developing end-uses for this waste.

First of all consider three major 'fibre waste streams' that we deal with:
1. Garment waste – 300 tonnes clothing collected per week (via charity shops)
2. Carpet waste – 2% of all landfill is carpet waste
3. Closed loop recycling, specifically between businesses.

How does this market look?

The first thing to say is that it doesn't look healthy and a personal opinion is that it is characterised by under-capitalised companies, many of which are leaving the market. Several well known names have pulled out in recent years and others have diversified into non-recycled areas for new growth. A number of small businesses have gone into receivership and as with much of the traditional textile industry in Yorkshire, some of the fibre re-pullers are exiting the market.

This is a shame for 'UK plc' because the government talks a lot about recycling. Even if the UK no longer manufactures large quantities of clothing, a garment or textile item from China, or other 'low cost' country, will still need to be disposed of after being worn or used in the UK. While government help is limited, the UK Department of Trade and Industry (DTI) has set up the Waste and Resources Action Programme (WRAP), although at present textiles is not a waste stream that it funds, concentrating on plastics, wood etc. WRAP has, however, supported a co-operative project between University of Bolton with University of Bradford to look at recycling carpet tiles.

In other countries, governments have taken a more active approaches to recycling for example; in Germany, the Association of Environmentally Favourable Carpets was formed with Euro5.5m support in 1990.

However, contact with a recycling expert at the DTI in London, indicated that DTI policy on recycling was driven by the EU agenda. Textiles are not on that EU agenda - an interesting insight with the way UK government is now working.

Why are companies in this market struggling in the UK?

A personal view suggests it is all down to the economics of the market and the low 'added value' products currently produced from re-cycled textile waste, coupled also with waste penalty costs. Landfill tax in the UK is reasonably low – £13/tonne against, for example, £45/tonne in the Netherlands. This is planned to increase to £35/tonne by 2012 and means that it only costs around £50/tonne to put waste into landfill in the UK. People therefore have no major incentive to recycle so we send 80% of our waste to landfill - in Switzerland the figure is 7% where different cost structures exist. This will change as the landfill tax and the bureaucracy of recycling increases.

RECYCLING STREAMS

Garment waste

Recycling garments is very well organised and the work of charities is significant with around 300 tonnes of clothing collected every week. The Salvation Army for example, run 1800 clothing banks and collect an average of 6.5 tonnes per bank per year. Only 10% by the way is sold in shops.

Much of this clothing waste is handled at the Oxfam waste-saver plant in Huddersfield, which processes 120 tonnes of material per week. 70% goes as second hand clothing – mainly to Africa and 25% is repulled in to fibre form or used as wipes.

A world leader in this area is a German company who sort all their clothes via a tiered selective system:
- Special clothing to theatrical hire companies for costumes
- 'Hip' clothing to boutique-type second-hand shops
- Normal second-hand clothing for charity shops and Third World areas
- The residue is pulled back in to fibre – often after sorting into fibre type / colours.

In the UK this is an innovative and well-developed, quasi-commercial market. Incidentally, one reason it works is that people tend to wash the clothes before taking them to the clothing banks – 'post consumer' carpet by contrast is often dirty, giving a different set of problems.

Carpet waste

To illustrate the challenges here, let us consider then for a minute recycling the carpet waste from a new hotel. An Anglo Felt customer, Brintons, is the supplier of the carpet and they may get up to 10% waste during fitting. It will cost them £30-£100/tonne to get that carpet back to a re-puller. This is because the carpet fitters will probably not have access to a baler so the carpet will be bulky to transport. Pulling back in to fibre form will cost around £120-£150/tonne. The problem is that the market price for low-grade waste is only £180-£200/tonne so the economics are not currently viable.

This is one reason why an innovative venture in Liverpool, UK failed. Carpet manufacturers in the Yorkshire/Lancashire area supported a company called WRACE Technology. Despite getting their carpet waste from the manufacturers delivered nicely baled, significant financial help from government agencies and state of the art Italian re-cycling machinery, this company failed – the economics just didn't seem to work.

Closed Loop recycling

Finally, let us consider business-to-business waste briefly where the market can operate successfully. Here, goods are generally presented to the waste merchant compressed so that transport costs are minimised. For example, fibre extrusion waste (from denier, colour changes etc) can be reprocessed and there is a ready market. It gets more complicated further downstream, however.

We have looked at reprocessing cotton gloves from car makers whose operators on the car assembly line typically get through 1 tonne of gloves per month. Anglo Felt has looked at reprocessing these back into sound insulation material in the car. This works fine in principle, but the problem was that the gloves needed cleaning before repulling. At present, the economics again do not stack up.

However, we have just started a co-operative project with Brintons, taking carpet edges from production in Kidderminster, re-pulling them back into fibre form and then putting them into carpet underlay. Because the waste is generally 80% wool or so, it is eminently suitable and provides good compression performance in the underlay and the product passes BS5808, the Industry standard.

How can companies develop and grow in this area?

A personal view, is that the key is designing and developing products to meet a technical specification for use in 'added value', technically demanding applications. This may be in a number of areas e.g.:
- Resiliency and compression recovery
- Flame retardancy.
- Sound insulation.
- Water holding.

Resiliency and compression recovery: Wool and wool/synthetic recycled fibre combinations can provide good recovery properties, which traditionally have been used in carpet underlay felts. Using modern foam bonding techniques, with high performance chemical binders, Anglo Felt has developed a new generation of underlay with enhanced properties whilst still retaining the proven properties of the traditional felt. This is the WoolSpring range which combine lightweight with easy-fitting together with excellent "under foot" performance, sound insulation and point-of-sale appearance for retail applications.

Flame retardancy: Flame retardancy (FR) is about using the natural retardant properties of re-cycled fibres (e.g. wool, hair) and perhaps improving them with specialist chemical treatments that are designed to enhance these inherent FR properties. In this area Anglo Felt has developed a sound insulation material that goes under theatre stages where Class 1 Building Regulations on flame retardancy apply and other more specialised materials for marine applications where FR properties are particularly demanding.

Sound insulation: In building / construction applications there are many opportunities for sound insulation, absorbent and transmission reduction materials. High loft materials from recycled fibres can provide excellent, cost effective performance.

In automotive, absorptive and anti-vibration properties are used more and more for additional performance and comfort in the vehicle. Recycled materials from renewable sources (e.g. wool, jute, hemp, flax etc.) are now accepted as high performance, cost-effective materials which can also help as the impact of the End-of-Life Vehicle Directive begins to be felt. Car makers want to use more recycled materials (some even use this in their advertising) and Anglo Felt are working to produce lighter weight products with higher acoustic performance using a systematic, scientific approach.

Water holding: A big area for us is using the natural water holding properties of wool and other natural fibres to produce a range of capillary mattings. These are extensively used in horticultural growing applications and in "Point-of-Sale" displays of flowers and shrubs in supermarkets and garden centres.

Our recently developed Algon capillary matting, a recycled fibre product with additional laminated surface layer to provide improved appearance and additional control of water release, is finding widespread use in many well known retail supermarket and garden centre outlets.

Appearance

Finally, mention must be made of appearance. It took five years to realise that even though the bulk of our products are out of sight, customers value how the product looks. We have discovered a way of foaming on colour on to our needle felts and can now offer a range of colours to meet our customers' requirements. This is useful to cover the range of coloured fibres, which inevitably arrive in the re-cycled fibre raw materials. An example is our WoolSpring range of bonded underlays, which meet the modern requirements for a light, easy-to-handle product that have visual impact, important in the retail underlay market.

Innovative products

Special products generally come from co-operation with like-minded, go-ahead customers and may often be relatively low volume but attractive margin, 'niche' products.

For example we use the following innovative effects:
- carbon impregnation for anti-static felts
- polypropylene blends and surfacing tissues for pre-finished mouldable parts
- coarse particle-coating to produce resilient underlays and mattings as well as sound absorbent and anti-vibration materials.
- Infra-red banks and specialist binders to produce:
  - tough anti-rub felts for the automotive industry,
  - anti-fatigue matting for warehouses and factories where tough disposable flooring options are required, and
  - water and oil absorbent matting, which have to be fork lift truck resistant!

Anglo Felt's motto is 'Innovating with Recycled Fibre' and at a time when many UK manufacturers are exiting this area, the company has invested more than £0.5 m, in the last 5 years, in additional manufacturing equipment. Basic raw materials are recycled fibres, both natural (wool, jute, cotton, hair etc.) and synthetic (polypropylene, acrylic, polyester) fibres. These materials are selected for product properties and performance e.g.

Flame Retardancy	- Wool, animal hair,
Resilience	- Wool, PP, acrylic, animal hair,
Water Holding	- Wool, cotton, jute
Sound Absorbency	- Wool, cotton, acrylic
Biodegradeability	- Wool, cotton, jute
Mouldability	- Polypropylene

Web-forming is selected from carding, cross-folding or air laying depending on end-use requirements. Bonding can by needling, thermal or chemical methods to provide the final properties and performance. Chemical bonding can be by full impregnation or one or two-sided foam bonding, which enables finishes or colour to be applied throughout the product or in a controlled layer on each side of the product. These techniques produce basic roll goods products with weights from 90 to 2000 g/m^2 and widths up to 4.5 m.

A range of after-treatments to confer additional properties is also available, these include impregnation, coating, infra-red singeing, calendering and lamination, to complete the manufacturing process.

To complement the roll goods manufacturing Anglo Felt has developed a significant slitting, sheeting and cut parts converting operation and distribution business in order to provide maximum service for the customer. This versatile operation has provided the possibilities for development of specialist technical products from recycled fibres.

Although there is significant 'in company know-how', many of the products we develop are designed in partnership with our customers where we can both benefit from our own specialist know-how and application knowledge. We are only a small company and there are many opportunities for co-operation with academic institutions where there is an enormous, untapped wealth of knowledge and ideas.

We have co-operated closely with Bolton, Bradford and Leeds Universities and we anticipate more of this type of co-operation in the future but we need to be involved at an early stage of the project.

Finally, the result of our focus on innovation over the last few years has been significant. Turn-over has increased by 50% and traditional, low value carpet underlay production, which was over 75% of turnover 10 years ago, is now less than 20% of our business. In its place Anglo now supplies a wide range of technical products for an increasing range of market applications.

CONCLUSION

In conclusion then, Anglo Felt have shown there is plenty that can be done with low cost, recycled fibres. However, the economics are often not very favourable at present and so the bad news for 'UK plc' is that plenty of companies are exiting the market. This is not really what we want – but the future may be brighter as progressive rises in landfill tax will make it more worthwhile for companies to consider recycling. Once people have to pay significant costs to have waste taken away and disposed of, then the market can begin to open up and develop. This will happen at some point, as environmental policy will gradually reduce the amount of material sent to landfill and Anglo Felt will be ready to meet this future challenge.

Part II

Sustainability and ecodesign

BUILDING ECODESIGN THROUGHOUT THE SUPPLY CHAIN: A NEW IMPERATIVE FOR THE TEXTILE & CLOTHING INDUSTRY

Tracy Bhamra

Department of Design & Technology, Loughborough University, Loughborough, Leicestershire, LE11 3TU, UK

ABSTRACT

In the electronic, electrical and automotive sectors, ecodesign is becoming a necessary and accepted addition to the product development process with products now appearing on the market having significantly improved environmental performance. The drivers for this have come from legislation, increasing customer awareness, a desire for an improved environmental image and the recognition that ecodesign can reduce costs for a company. However, to date there is little evidence of a similar move towards ecodesign within the textile industry despite many supply chain members having clear environmental credentials. This paper will highlight some of the key drivers for ecodesign within the textile and clothing industry illustrating the learning that can be transferred from the other sectors adopting this approach to product development. It will introduce some of the methods that can be employed to encourage ecodesign helping to ensure that in future products from this sector illustrate clear environmental improvements to customers. With the complexity of the structure of the textile and clothing industry and the nature of environmental problems it is particularly important that all supply chain members work together to ensure that ecodesign becomes an accepted part of good product development. Such a proactive approach could result in a significant improvement of image with the consumer.

INTRODUCTION

Ecodesign as a subject has developed considerably over the last ten years with the increase in awareness of environmental issues in industry [1, 2, 3]. Until recently, the usual response to environmental problems was to reduce pollution and waste after it had been produced. Attention then moved away from these 'end-of-pipe' approaches to 'cleaner' manufacturing which results in less waste and pollution being generated. There was then the realisation that major environmental impacts arise from the material choices and from the use and disposal of products [4].

The most advanced companies are now moving beyond the compliance mentality and being proactive in shaping future markets, consumer needs and influencing legislative developments. They see environment as an opportunity rather than a threat, recognise that 'prevention is better than cure' and are attempting to 'design out' rather than simply manage the problems.

Ecodesign is understood to be the systematic integration of environmental considerations into the design process across the product life cycle, from cradle to grave [5, 6, 7]. This approach aims to reduce and balance the adverse impact of a manufactured product on the environment by considering the product's whole life cycle (illustrated in figure 1 below) from raw materials acquisition, through manufacturing, distribution and use to final recycling and disposal.

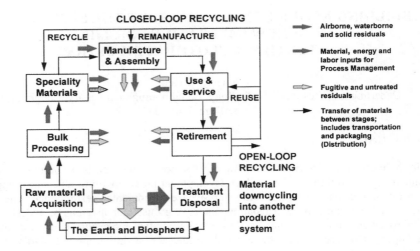

Figure 1 Product Life Cycle System [8]

There are various influences, both internal and external driving organisations towards implementing ecodesign. These are summarised as:

- Cost savings: integrating environmental issues into product development can result in cost savings such as less raw materials used, less waste produced, energy efficiency and water efficiency [9].
- Legislative regulations: these are becoming more and more important to companies as they are increasing in both the country in which they operate and to which they export [9].
- Competition: pioneering companies have realised that they may gain some competitive advantage by considering ecodesign [10, 11].
- Market pressure: Ecodesign can be an effective way to improve an organisation's environmental performance and therefore help to meet the increasing market pressure associated with the environment [9].
- Industrial customer requirements: many suppliers are now being asked to meet their customer's environmental requirements[11].
- Innovation: new market opportunities can be opened up by integrating environmental issues into product development as this stimulates product and process innovation [10].
- Employee motivation: introducing environmental considerations in a organisations is an effective way to engage and motivate employees as they can actively contribute and improve their working environment [11].
- Company responsibility: many more companies are becoming more aware of their responsibility towards the environment and the role they need to play in a more sustainable society [11].
- Communications: many organisations are using the environment as an effective communication tool with all stakeholders. It is providing a new mechanism to promote both the organisation and their products or services [7].

This mix of internal and external pressure has resulted in a real drive in many industrial sectors to consider ecodesign as an integrated part of all product development.

BACKGROUND TO ECODESIGN

The concept and practice of ecodesign is the response from the design community to increasing environmental pressures, limits and awareness. In the late 1980's and early 1990's, coinciding with the "Green Consumer" revolution [12,13], design began to be viewed as important in the development and launch of more mainstream eco-products and in enhancing consumer acceptance of these. This also helped portray an environmental profile for companies whilst applying and interpreting various new and clean technologies [14].

Definitions and descriptions of ecodesign

The last two decades have seen a proliferation of terminology relating to the incorporation of environmental considerations into design. This has included ecodesign [15]; environmentally conscious design [16]; Design for the Environment [11,17]; Life Cycle Design [18]; EcoRedesign [19]; and green, ecodesign and sustainable design [20]. Some environmentally related design terms come from Dewberry and Goggin [20]. They propose three ecodesign approaches as: green design; ecodesign; and sustainable design:

- *Green design:* has a single-issue focus, perhaps incorporating the use of some new material, such as recycled or recyclable plastic, or consider energy consumption.
- *Ecodesign:* adopts the lifecycle approach, exploring and tackling all or the greatest impacts across the products lifecycle.

Models are used to explain ecodesign in more detail and one by Brezet [21] provides a clear outline of the key design criteria and consideration. This model proposes a four-step model of ecodesign innovation and is illustrated in figure 1 below. These steps are described as:

- *Product improvement:* The improvement of existing products with regards to pollution prevention and environmental care. Products are made compliant.
- *Product redesign:* The product concept stays the same, but parts of the product are developed further or replaced by others. Typical aims are increased reuse of spare parts and raw materials, or minimising the energy use at several stages in the product life cycle.
- *Function innovation:* Involves changing the way the function is fulfilled. Examples include a move from paper-based information exchange to e-mail, or private cars to 'call-a-car' systems.
- *System innovation:* New products and services arise requiring changes in the related infrastructure and organisations. For example, a changeover in agriculture to industry-based food production, or changes in organisation, transportation and labour based on information technology are typical.

To move from level 1 to level 4, increasing amounts of time and complexity are required, which leads to higher eco-efficiency improvements. This model suggests that these more complex ecodesign innovations will (or can) only be achieved over a significant time period, say 10-20 years.

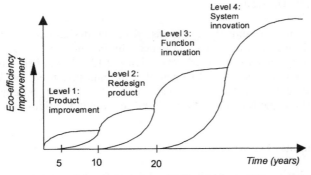

Figure 1 Four Stage model of ecodesign innovation [21].

TEXTILE & CLOTHING SUPPLY CHAIN AND THE ENVIRONMENT

It has been recognised that improving the communication and understanding between the various stakeholders within the textile and clothing sector is of paramount importance [22]. Connections have been made between communication and improved environmental performance and these suggest that ecodesigned textile and clothing products cannot be successfully developed until a deeper understanding of the supply chain is attained [22]. Designers are often isolated, for example, from the finishers and therefore unaware of the environmental demands being made to this section of the product development process. A summary of existing communication channels throughout the textile industry have been illustrated in figure 2 below.

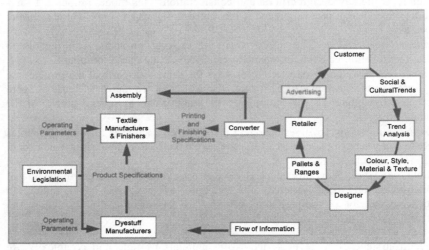

Figure 2 Summary of communication channels throughout the textile & clothing industry [22]

Figure 2 highlights the isolation of the designer and also emphasises how the designer, consumer, trend analysis and partly the retailer are divorced from the impacts of environmental legislation impacting upon operation.

Increasing environmental awareness

Within the sector and at all stages of the supply chain there has been a significant increase in environmental awareness and this is reflected in the number of companies now undertaking environmental reporting which includes setting and making public targets for improvement. This is particularly evident within retailers who are also encouraging members of there supply chain to take a similar approach [23]. However there is still very little evidence of any significant activity which takes into account the design of the products. This is surprising given that in 1997 a survey of 70 UK textile and clothing designers found that 80% of the respondents recognised that there was an increasing demand for improved environmental products [24]. Respondents of this survey also identified a number of barriers which they felt inhibited the adoption of more environmentally responsible strategies. The main ones involved:

- lack of environmental awareness and appropriate environmental information;
- lack of control over types of materials and processes used;
- lack of senior management commitment;
- concern over increased costs; and
- lack of time.

The survey also revealed that the designers perceive consumer demand to be the main driving force for improved environmental demand which at the present time appears still to be low. However there is evidence that up to 60% of the population in the UK are keen to buy environmentally improved products [25], but often struggle to find products that could be viewed as such. The main reason is that very few mainstream textile and clothing products are being designed with the environment in mind. In addition many products do not provide enough information in order to assess effectively their environmental performance. Clarifying issues to consumers through the use of care labels which identify clearly the ecological efficiency of products is necessary if any improvement is to be made.

To date, the designer has largely been perceived as the weak link in the chain of environmental improvements [26] as they are often remote from the consequences of their design specifications which may impact significantly on the environment. Manufacturers also argue that the designer's promotion of the "Ecology Look" in the late '80s and early '90s, promoting natural fabrics in earthy tones largely confused or hindered any real environmental improvements. This was because little if any consideration had been given to how the yarns, fabrics and colours were made, and the environmental impact these processes may have had.

Watson [27] also argues that the notion of 'green textiles' has trivialised the problems manufacturers have to deal with because they appear to provide instant answers, although these answers are confused. The small scale benefits of these products are being promoted when the industry has got to deal with these problems on a mass scale. Watson [27] believes that demands for the most 'environmentally friendly' fibre are unrealistic, whilst demands for the most 'sustainable process' would be more productive in improving the environmental performance of the production and processing of textile and clothing products. In the long run this will require producers and retailers, along with the designers, to work together.

ECODESIGN IN THE TEXTILE AND CLOTHING SECTOR

Although there is currently little evidence of textile & clothing designers being encouraged to consider ecodesign there is a rise in awareness both within the sector and amongst consumers. Therefore the opportunity exists for ecodesign to be implemented in a systematic fashion with the supply chain working in partnership to make it happen.

The early part of this paper demonstrated that when thinking about ecodesign consideration of the whole lifecycle of the product was key. Within the textile and clothing sectors it is only possible to consider the whole life cycle if worked from a supply chain perspective. However, in order to identify the key areas for ecodesign to focus on it is necessary to have a detailed understanding of the environmental impact of the particular product to be redesigned. There have been some studies of garments that have examined environmental impacts over the whole lifecycle and these can be a good starting point when drawing up ecodesign guidance for designers.

In their study of a woman's knit polyester blouse, Franklin Associates [28] concluded that the manufacture of the product did not have the most significant environmental impacts; these arose from the use of the product. This result is due to the laundering of the product which uses significant amounts of energy, water and detergent in its washing and ironing. Therefore one of the key considerations for designers is to ensure that these products require less of this type of 'maintenance'. However, an alterative view would be that appliance and detergent manufacturers need to improve the design of their products.

In their study of two products for Marks & Spencer, ERM [29] concluded that between 76% and 80% of the energy consumption of the garment is from the use stage with around 13% coming from the manufacture of the product itself. This again provides designers with clear targets for redesign. The problem with both these studies is that they do not highlight any environmental problem that could be occurring at the raw materials or production stages of the life cycle and therefore it becomes very difficult to enable designers to design out these environmental problems.

Brown and Williams [30] however present a useful checklist for an "ideal" garment based on their experience within Patagonia which covers all stages of the life cycle and therefore provides designers with more direction. This checklist approach is useful however the way in which design within the sector is currently organised (see Figure 1) and is likely to make it difficult for designer to both understand the particular environmental problem and to find a way in which they can affect it. For example one item in production part of the checklists identifies zero-waste as a target. This may be difficult for the designer to address but could be incorporated into the customer's purchasing policy.

One organisation that has managed to incorporate some elements of ecodesign into their products is Interface Fabrics [31]. This company has made a range of substantial environmental improvements through the implementation of an Environmental Management Systems (EMS) and one of these was to develop a new "eco-product" every year. This has led to a number of products manufactured from 100% recycled materials such as polyester fibre produced from post-consumer PET bottles and woollen products made from reclaimed Ministry of Defence sweaters and hosiery [31]. Other areas of improvement that have been influenced by ecodesign is packaging were many opportunities for re-use of materials were identified and have been introduced [31].

46

CONCLUSIONS

The first part of this paper outlined the experience in ecodesign gained within other sectors of the last twelve years. It has taken time for companies to understand the subject, identify how it can be implemented and to begin to develop suitable methods and tools for designers so that ecodesign can become an integral part of their designing. Consequently, there are now product and services on the market that fit within all levels of Brezet's model [21], this is now showing that higher levels of eco-efficiency can be achieved.

The textile and clothing sectors however appears to be a long way from this. There is little evidence of companies integrating ecodesign in any way at all and very few examples of products on the market where this has been a core consideration in design. As illustrated earlier it is believed this is due to the complexity of the supply chain and the relative isolation of some designers from the supply chain. Despite these reasons this cannot continue as customers are becoming more aware of environmental issues and even beginning to base purchasing decisions on these considerations. In addition, the electrical, electronic and automotive sectors now face 'Producer Responsibility' Legislation, making them responsible for their products at the end of life, how long will it be before the textile and clothing sector faces similar legislation? The sector will be ill-prepared when this comes along.

The few examples of ecodesign that currently exist from the sectors are clearly in the first two levels of Brezet's model [21], product improvement and product redesign, where eco-efficiency gains are at their lowest. The sector therefore has a long way to go to catch up with the levels of eco-innovation being achieved elsewhere. There is clearly a lack of tools and methods available for designers for the textile and clothing sector which may be hindering its implementation but this lack of development is likely to be due to lack of demand. If companies from the sector required these tools for their designers, they would become available.

As outlined earlier in this paper it is imperative that all the supply chain impacts are understood by designers and they then find ways to design out these environmental problems. Consequently opportunities now exist for research partnerships between academia and the sector in order to better understand the impacts and develop suitable tools and methods for designers. In addition, as textiles and clothing are part of a complex chain which also include maintenance (i.e. laundering), new partnerships must emerge between the sector, appliance manufacturers and detergent manufacturers if environmental impacts are to be significantly reduced.

REFERENCES

1   E Von Weiszacker, L H Lovins and A B Lovins, *Factor Four: Doubling wealth, halving resource use*, London, Earthscan, 1997.

2   P Hawken, *The Ecology of Commerce*, London, Weidenfeld & Nicholson, 1993.

3   M Wackernagel and W Rees, *Our Ecological Footprint: Reducing Human Impact on the Environment*, USA, New Society Publishers, 1996.

4   R Roy, 'Sustainable Product-Service Systems', *Futures*, 2000, **32**, (April), 289-299.

5   T E Graedel and B R Allenby, *Industrial Ecology*, New Jersey, Prentice Hall, 1995.

6   E Dewberry, *Ecodesign - Present Attitudes and Future Directions: Studies of UK Company and Design Consultancy Practice*, Open University, PhD thesis, 1996.

7   M Simon, S Poole, A Sweatman, S Evans, T Bhamra and T McAloone, 'Environmental Priorities in Strategic Product Development', *Business Strategy and the Environment*, 2000, **9**(6), 367-377.

8   G Keoleian and D Menerey, *Life Cycle Design Guidance Manual: Environmental Requirements and the Product System*, USA, Environmental Protection Agency (EPA), 1993.

9   F O'Connor and D Hawkes. 'A Multi-Stakeholder Abridged Environmentally Conscious Design Approach', *The J of Sustainable Product Design*, 2001, **1**, 247-262.

10   F Rubik, 'Environmental Sound Product Innovation and Integrated Product Policy', *The J of Sustainable Product Design*, 2001, **1**, 219-232.

11   R Sroufe, S Curkovic, F Montabon and S A Melnyk, 'The New product Design Process and Design for Environment': "Crossing the Chasm", *Intl J of Operation and Production Management,* 2000, **20** (2), 267-291.

12   J Elkington and J Hailes, *The Green Consumer Guide*, London, Gollancz, 1987.

13   J Elkington and J Hailes, *Manual 2000*, London, Hodder and Stoughton, 1998.

14   C Ryan, M Hosken and D Greens, 'Ecodesign: design and the response to the greening of international markets'. *Design Studies*, 1992, **13**(1), 3-22.

15   Eco2-irn, *Defining Ecodesign*: minutes from quarterly meeting of the Economical and Ecological Sound Design and Manufacture - Interdisciplinary Research Network (Eco2-irn) workshop, Manchester Metropolitan University, Forum No. 13, Feb. 1994.

16   T McAloone, *Industry Experiences of Environmentally Conscious Design Integration: An Exploratory Study*, Cranfield University, PhD Thesis, 1998.

17   C van Hemel, *Ecodesign empirically explored – Design for Environment in Dutch Small and Medium Sized Enterprises,* Delft University of Technology, PhD Thesis, 1998.

18   K A Keoleian and M Menerey, 'Sustainable Development by Design: Review of Life Cycle Design and Related Approaches', *Air & Waste*, 1994, **44** (May) 645-668.

19   C Ryan, 'From EcoRedesign to Ecodesign', *Ecodesign*, 1996, **4**(1), 5-7.

20   E Dewberry, and   P Goggin, 'Spaceship Ecodesign', *Co-Design: the interdisciplinary journal of design and contextual studies,* 1996, **05 06**(01 02 03), 12-17.

21   H Brezet, 'Dynamics in ecodesign practice'. *UNEP Industry and Environment* 1997 **20**(1-2) 21-24.

22   DEMOS *Textile Finishers Association: Executive Summary,* UK, DTI Publication, 1994.

23   N Robin and L M Humphrey, *Sustaining the Rag Trade,* International Institute for Environment & Development, April, ISSN 1562 3319, 2000.

24   ETBPP, *Water Chemical use in the Textile Dyeing and Finishing Industry,* Good Practice Guide GG62, ETBPP, 1997.

25   ETBPP, 'Cost effective Environmental Improvements in the Textile Industry Through Design' *Textiles, Design & Environment: Online Conference,* Available at http://www.cfsd.org.uk/on-line-tde/etbpp.htm, 1997

26   National Consumer Council, *Consumers and the Environment: Can Consumers save the planet?,* National Consumer Council, September, PD 38/B2/97, 1997.

27   J Watson, *Textiles and the Environment.* Economist Intelligence Unit, Special report, 1991.

28   Franklin Associates, *Life Cycle Analysis(LCA): Woman's Knit Polyester Blouse,* 1993, [online] available at http://www.fibersource.com/f-tutor/LCA-page.htm (23rd March 2004).

29   Environmental Resources Management, *Streamlined Life Cycle Assessment of Two Marks & Spencer plc Apparel Products,* 2002   [online] available at http://www2.marksandspencer.com/thecompany/ourcommitmenttosociety/environme nt/pdfs/Final_LCA_report.pdf (23rd March 2004).

30   M S Brown and E R Williams, 'Quick & Dirty Environmental Analyses for Garments: What do we need to know' *Journal of Sustainable Product Design,* 1 (April), 28-35

31   ETBPP, *Sustainability Drives Waste Reduction,* ETBPP, June, Case History CH267, 2000.

SUPPLY CHAIN PARTNERSHIPS FOR SUSTAINABLE TEXTILE PRODUCTION

John R Easton
R&D Ecology Dept, DyStar Textilfarben GmbH & Co Ltd

INTRODUCTION

In the latter half of the 20th Century company director's performance and bonuses were assessed on the basis of whether they had succeeded in building "Long Term Shareholder Value". But financial mismanagement as evidenced by major corporations such as Enron, Worldcom, and more recently Shell, has given weight to those who argue that companies, especially multinational companies, and their directors, have a much wider responsibility than simply to shareholders.

The term the Triple Bottom Line was coined to describe company accountability not only in financial terms to shareholders and investors but also in social terms e.g. acceptance by the communities in which the firm operates, and in environmental terms by assessing the impact on the environment of the firm's own activities, or the activities of the manufacturing chain which supplies the products that the firm sells.

Consumer expectations of company behaviour have increased and not only in the rich G8 countries. In a millennium poll in 23 countries 60% of 25,000 consumers said they expected businesses to tackle the issues of fair labour practices, business ethics and environmental degradation in addition to delivering profits and jobs

So how does this affect the clothing and textile industry? In their book entitled "Sustaining the Rag Trade" published in 2000 by IIED, Roberts and Humphries identified four key elements of the sustainability of the clothing supply chain. These were:

- Eliminating environmental hazards
- Improving energy and water efficiency
- Cutting pollution and waste
- Establishing social justice

In addressing the wet processing stage of the clothing supply chain the issues shown in Table 1 were identified by the authors under each of the four headings.

Table 1. Key Sustainability Issues in Dyeing & Finishing

Environmental Hazards	Carcinogenic, toxic and allergenic dyes
	Use of PCP and formaldehyde in finishing
Materials Efficiency	High water consumption for dyeing
	High energy use for dyeing and finishing
Pollution and Waste	Biodegradation of surfactants
	Emissions from chlorine bleaching
	Effluent from unfixed dyes
Social Justice and Equity	Child labour, job security, equal opportunity, wages, working hours, freedom of association, collective bargaining

Careful selection of dyes and chemicals can play a decisive role in addressing the first three of these issues (the social justice issues are outside the scope of this paper).

Until very recently, companies have tended to address social and environmental issues separately although the Corporate Responsibility (CR) paradigm is beginning to bring these two strands closer together both conceptually and organisationally in many leading companies. In the apparel and clothing supply chains the social concerns have been primarily addressed at the level of the garment suppliers through individual company Codes of Conduct, or cross-sector codes such as the Ethical Trading Initiative, or through standards such as AA1000 or SA 8000.

The focus of companies environmental concerns has widened from the companies' own activities e.g. distribution, packaging waste, energy efficiency to that of the environmental impact of the product supply chain and the potential for chemical contamination in the final consumer article. From this viewpoint the important developments have occurred in Environmental Management Systems (EMS) and ecolabel schemes, as well as company-specific Restricted Substances Lists (RSLs).

We can now go on to look at the key environmental drivers affecting the textile supply chain.

KEY ENVIRONMENTAL DRIVERS IN THE TEXTILE INDUSTRY

Legislation

There are two main types of legislation which impact on the textile industry and the dyestuff and chemical companies who supply it:
• chemical control legislation and
• pollution control legislation.

Chemical control legislation affects the innovation, classification and labeling, supply and use of textile dyes. The most notable pieces of European legislation are the Dangerous Substances Directive (67/548/EEC), the Dangerous Preparations Directive (1999/45/EC), and the Marketing & Use Directive (76/769/EEC) and their subsequent adaptations to technical progress (ATPs) and amendments. In the United States similar controls exist through the Toxic Substances Control Act (TSCA).

In October 2003, after a period of internet-based public consultation the European Commission published its proposals for a major revision of chemical legislation in Europe. The new chemical policy is known as REACH (the Registration, Evaluation, and Authorisation of Chemicals) and is expected to come into force across the enlarged EU sometime in 2006.

The other major area of environmental legislation impacting on the textile industry is pollution control legislation which affects the use and disposal of textile dyes as well as addressing water and energy consumption, air pollution, wastewater treatment and waste disposal. The most important items of European legislation in this area are the IPPC Directive, the Urban Wastewater Treatment Directive and the Water Framework Directive.

Azo dye legislation

On the 15th July 1994 the German government enacted legislation effectively banning the use on certain consumer goods of azo dyes, which could undergo reductive cleavage of the azo bond(s) to release any of 20 specified aromatic amines which were known or suspected carcinogens.

The EU harmonising legislation, the 19th Amendment to the Marketing and Use Directive (azocolourants) (2002/61/EC), came into force in 2003. The major European dyestuff manufacturers no longer produce dyes affected by this legislation eg benzidine-based direct dyes which were found to be carcinogenic over 30 years ago. However a large number of smaller manufacturers in Asia still produce these dyes.

Ecolabels

Over the last 10 years the development of various eco-labelling schemes, designed to assure consumers of the safety and environmental acceptability of a particular product, has raised the profile of environmental issues within the textile chain. The most successful of the textile eco-labelling schemes has been the Oeko-tex Standard 100 scheme run by the independent "International Association for Research and Testing in the Field of Textile Ecology" and operated by franchised testing institut in Europe, USA and Asia. Over 6000 companies worldwide have been awarded Oeko-Tex Standard 100 certfication.

By comparison the official EU scheme for textiles has been largely ineffective due to its complexity and cost. Only 53 EU ecolabels have been awarded for textile products since the criteria were established in 1999.

Oeko-tex Standard 100 is in fact a "Human Ecology" label as it is only concerned with possible negative effects of textiles on the health of the wearer. The latest version of the Standard has four basic product categories:
- Babywear
- Textiles in contact with the skin
- Textiles not in contact with the skin
- Household textiles

and limit values for chemical residues and fastness criteria are laid down for each of the product categories. The specification for babywear is the most demanding. The parameters most affected by the coloration stage, and therefore critically affected by the choice of dyes and chemicals, are those covering heavy metals, carcinogenic dyes, allergenic dyes and colour fastness.

Environmental Management Systems (EMS)

Alongside the development of product-related ecolabels there has been a parallel growth throughout the 1990s of Environmental Management Systems which focus more on how a company manages the environmental impact of its activities as a whole rather than on the detailed assessment of individual products.

The British Standards Institute (BSI) launched the world's first Environmental Management Standard BS 7750 in 1992. Shortly after the introduction of BS 7750 the European Commission created the Eco-Management Audit Scheme (EMAS) by Regulation No 1863 in 1993. EMAS is based on the same principles as BS 7750 but additionally requires the publication of statement of environmental performance that has to be independently verified.

The international standards body ISO responded to the lead of BSI and others by developing an equivalent international standard ISO 14001 in 1996. In the same way that ISO 9001 is meant to signify a quality-oriented organisation so ISO 14001 is intended to mark out companies with a systematic approach to environmental management and a commitment to reducing the environmental impact of their operations. Recently there has been a rapid uptake of ISO14001 by manufacturers in

Asia as a means of assuring European and N. American procurement managers of the environmental probity of their operations.

A common feature of all the standards is the emphasis on external audit and accreditation by authorised bodies. Furthermore, ISO 14001 is also an effective mechanism for transmitting environmental issues along the supply chain as it requires an organisation to question its suppliers on their environmental performance and in best practice companies leads to a dialogue on key environmental issues affecting the vendor – buyer relationship.

Retailers

Several leading clothing suppliers and retailers in Germany were involved in the early discussions about textile ecolabels and started to include product safety or human ecology specifications into their supplier manuals from the late 1980's onwards. Companies such as Quelle, Otto, Steilmann and Karstadt & Neckarmann developed standards or eco-specifications similar to that of the fledgling Oeko-tex 100. This trend was given further impetus in 1994 by the introduction in Germany of the amendment to the Consumer Goods Legislation which banned the use of certain azo dyes.

From the early 1990s onwards, starting in Germany and Scandinavia, the mail order houses and clothing retailers began imposing chemical and environmental restrictions on suppliers in order ensure product safety standards and address the increasing concern of consumers for environmentally friendly goods.

The approach of the leading UK retailer Marks & Spencer was rather different to that of continental firms in that it introduced an Environmental Code of Practice in 1995 encompassing the whole of its textiles business not just for special eco-brands or lines.

In contrast C&A adopted the Oeko-tex ecolabel as its corporate standard and began rolling it out across various products in-store. These are identified as "safer clothing" options for the consumer by the use of an additional swing ticket bearing the Oeko-tex slogan "Confidence in Textiles" and the Oeko-tex logo. It has to be said that since the withdrawal of C&A from the UK in 2000 the Oeko-tex label is now virtually invisible on British High Streets.

Green Consumers

The 1990s have been dubbed the green decade by some commentators. The speed at which the environment has become a central commercial issue has left many businesses and industries unsure of their response. There has been tremendous growth in membership of green groups in the UK from the archly conservative eg the National Trust whose membership doubled from 1 million to 2 million in the 90s to the more radical groups such as Greenpeace and Friends of the Earth. At the same time there has been a huge expansion in environmental consultancy and environmental technology businesses as new opportunities opened up for companies previously confined to the relatively unglamorous waste disposal industry.

But perhaps the greatest revolution has been in public awareness and concern for the natural environment and the willingness of consumers to exercise a choice in favour of products that don't "cost the earth". In the UK context, an influential book in this greening process was the publication in 1998 of The Green Consumer Guide which became a popular best-seller. The following year the Green Party gained an astonishing 15% of the vote in the European Parliament Elections.

Each year since 1988 the well-respected polling organisation MORI has conducted an annual survey of Business and the Environment and their report includes a description of a typical British green consumer eg
- Regularly uses a bottle bank
- Take own waste paper for recycling
- Buys organically grown food
- Uses unleaded petrol
- Avoids products of companies with poor environmental record

Its 1997 report identified 37% of the population as green consumer activists based on a sample of 2000 adults. Another survey in 1995 identified different types of green consumerism in the UK.

Green Consumers

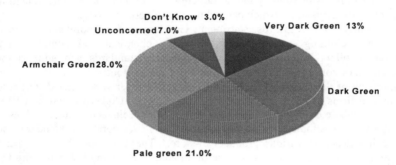

Don't Know 3.0%
Unconcerned 7.0%
Very Dark Green 13%
Armchair Green 28.0%
Dark Green
Pale green 21.0%

Armchair Greens
These people state that they are concerned about the environment but this concern does not carry through into their lifestyle or purchasing behaviour

Pale Greens
This group is concerned about the environment and if offered a choice between two similar products at the same price will opt for the one that is marketed as environmentally-friendly

Dark Greens
These people are concerned about the environment and are willing to pay a premium for goods and services that they believe have a reduced environmental impact compared to conventional products.

Very Dark Greens
These people will only buy what they consider to be natural or organic products and are often to be found protesting against motorway extensions, nuclear installations, GM crops etc.

Pressure groups

The membership of environmental pressure groups and NGOs has grown rapidly in the last 10 years. The more activist groups target high profile multinational companies in order to challenge them not only on the environmental impact of their own operations but also that of their suppliers.

The focus of the Clean Clothes Campaign is more on social conditions in the garment and footwear manufacturing factories supplying famous brands such as Nike, Gap, Adidas etc. Such groups are becoming more influential as "Corporate Social Responsibility" and "Sustainable Development" move up the agenda of the large multinational companies.

Ethical Corporation Magazine has organized Conferences on CSR in Asia, Europe and N. America addressing the issues of CSR in the Supply Chain and leading companies have been quick to use the opportunity to present themselves as leaders in the field e.g. Hennes & Mauritz, Marks & Spencer, and Nike in the apparel area. Another indicator is that membership by major retailers of the Ethical Trading Initiative has increased steadily in the last few years with Gap Inc joining recently. Similar organizations have been established in the US e.g. Fair Labor Association, etc.

Friends of the Earth has recently published its second report on how retailers in the UK rate on managing risky chemicals in their supply chains. It is no surprise to find companies such as Ikea and Marks & Spencer are near the top of the league table as they have been managing these issues for a decade or more and have the added advantage of dealing exclusively in own-brand products.

Greening the textile and clothing supply chain

The textile supply chain is a classic example of a "Buyer-Driven Global Commodity Chain" (Gereffi, 1994) in which it is the large retail groups and brand name marketers in N.America and Western Europe who exercise the greatest influence on the specification of the goods they source, increasingly from the Asia Pacific region.

Due to the globalisation of the textile supply chain, and consumer/NGO pressure on "Corporate Responsibility" in Western markets, environmental drivers have been communicated to all the major textile and clothing producing regions of the world.

So even if an individual producer country does not have tough pollution control legislation or liberal labour laws, minimum standards of working conditions and safety, health and environmental requirements may be imposed by the European or North American purchasing organisation.

What is the role of the dyestuff supplier in "Greening the Textile Supply Chain"?

An initiative adopted by the chemical industry internationally termed "Product Stewardship" (as part of the Responsible Care programme) is based on the idea that the manufacturer's responsibility for the product does not end when the delivery truck leaves the factory gate. It requires suppliers to be proactive in providing information to customers on the health, safety and environmental impacts of their products in the context of their storage, use and disposal. Environmental responsibility should characterise all the major functional activities of the company and extend across the whole life-cycle of the product from "cradle to grave".

In DyStar we have developed a coloration ecology concept which recognises the environmental influences acting on our customers and seeks to build the most important issues and concerns into the key business processes of innovation, manufacturing and sales & marketing in order to deliver new products, application processes, technical service and product information which meets the customer's needs.

Controlled coloration

There are two basic approaches to pollution control
- effluent treatment - or end-of-pipe solutions
- waste minimization - or source reduction solutions

The first approach has no financial payback and is literally money down the drain. The Waste Minimisation approach not only reduces environmental impact but also delivers reduced costs - the so-called "Win-Win Scenario".

Dyestuff companies generally do not supply effluent treatment solutions although they may offer general advice on the treatability of their products. DyStar's expertise is in the invention, manufacture and application of textile dyes. Our pollution control solution is either a waste minimisation, or a clean production, solution based on the concept of Controlled Coloration. The way we deliver this is by environmentally-oriented product and process innovation.

Controlled coloration permits textile coloration processes to be carried out in a way which minimizes impact on the environment . The controls which the dyestuff manufacturer can exert are:
- Control of dyeing behaviour
- Control of product quality
- Control of application processes
- Control of environmental impact

For example some of the factors which must be taken into account when designing reactive dyes for reduced environmental impact are:
- Careful choice of intermediates – no banned amines, minimum AOX
- High colour yield – high fixation multifunctional dyes leading to reduced levels of colour in effluent
- Suitability for ultra low liquor ratio dyeing machinery – to minimize energy, water and chemicals consumption
- "Right First Time" dyeing through dyestuff compatibility – to minimise wasteful shading additions or reprocessing

Eco-Innovation (making products cleaner by design) has a major role to play in "Greening the Supply Chain" and ensuring that a greater degree of resource efficiency or resource productivity is available to the textile wet processing sector. The environmental problems facing the textile wet processing industry cannot be solved using outdated products, processes or machinery. Innovation is required to address the environmental issues facing the supply chain and dyestuff manufacturers have a key role to play.

Using a Life Cycle Assessment approach within our coloration ecology concept we have developed a number of new product ranges to address the environmental issues affecting particular fibre dyeing segments of the industry.
Some examples of recent DyStar innovations are
- High fixation reactive dyes for reduced dyeing cycles and minimum colour in effluent
- Novel disperse dye chemistry to achieve high fastness without the need for reductive clearing
- Metal-free reactive dyes for wool – an environmentally-friendly alternative to chrome dyes

However one example of environmentally-oriented product innovation deserves special mention - the Procion XL+ dyes.

Let's just look at what one of the leading business thinkers has to say about the relationship between innovation and environmental improvement. Professor Michael Porter of Harvard Business School has written influential books on Competitive Advantage, not only between companies but also between nations. In an article in 1995 he underlined the need for innovation to raise "resource productivity":

"Increasingly the nations and companies that are the most competitive are not those with access to lowest-cost inputs but those that employ the most advanced technology and methods in using their inputs...environmental progress demands that companies innovate to raise resource productivity - and that is precisely what the new challenges of global competition demand"

DyStar's Procion XL+ dyes were designed to do just that by permitting shorter dyeing cycles on cellulose textiles (minimum prepare or scour dye methods).This leads to enhanced productivity and reduced water and energy consumption. As a consequence, DyStar has won two prestigious awards for the development of this dyestuff chemistry and the application processes it facilitates. In order to deliver environmentally-oriented process innovation we work closely with leading textile machinery and equipment suppliers. Examples of optimised application processes developed in partnership with third party technology suppliers were showcased on the DyStar stand in 2003 at the International Textile Machinery exhibition at the NEC, Birmingham

CONCLUSION

The combined effect of the drivers referred to in the preceding discussion has been to raise the profile of environmental issues within the industry – a process which has been described as "Greening the Textile Supply Chain". Due to the globalisation of the textile supply chain these environmental drivers have been communicated to all major textile and clothing producing regions of the world.

At DyStar we believe that globally co-ordinated supply chain partnerships operating at a local level are essential for the realization of sustainable textile production. Only by all the actors in the textile supply chain working together can environmental improvement be promoted through the introduction of ecologically-optimised products and cleaner production techniques.

By establishing such co-operative supply chain partnerships the industry will be better equipped to respond to the environmental challenges it will face in the 21st century.

© John Easton 2004

REFERENCES

1 G Gereffi, *The Organisation of Buyer-Driven Global Commodity Chains, in Commodity Chains and Global Capitalism*, eds G.Gereffi and M.Korzeniewicz, Praeger, Westport CT, Chapter 5, 1972.

2 K Dickerson, *Textiles and Apparel in the Global Economy*, 3rd Edition, Pearson New York, 1998.

3.   P Dicken, *Global Shift: Transforming the World Economy*, 3rd Edition, Paul Chapman Publishing, London, 1998.

4.   N Robins, L Humphrey, *Sustaining the Rag Trade*, IIED, London, 2000.

5. M Porter and C van der Linde, "Green and Competitive: Ending the Stalemate", *Harvard Business Review*, Sep/Oct 1995, p.120.

MOLECULAR MANUFACTURING FOR CLEAN, LOW COST TEXTILE PRODUCTION

David R. Forrest,
Naval Surface Warfare Center, West Bethesda, Maryland USA and
Institute for Molecular Manufacturing, Los Altos, California USA

ABSTRACT

Molecular manufacturing is an emerging technology that is being developed to build large objects to atomic precision, quickly and cheaply, with virtually no defects. When it matures, it promises to be an energy efficient and environmentally benign way to make textiles and textile products.

Current efforts to apply nanotechnology to textiles promise exciting innovations such as:

- Lightweight nanotube fibers stronger than steel
- Fabrics able to sense temperature and control their breathability
- Clothing able to sense injury and provided immediate delivery of medication to a wounded soldier

However, molecular manufacturing will provide atomic control over the structure of a fabric, promising revolutionary changes far beyond current advances:

- A garment could programmably increase or decrease its own size as needed
- A fabric could change its own color and patterns on demand
- The breathability of a textile could be variable and self-regulating
- By integrating molecular robotic components into the material, a fabric could be made to be self-cleaning and self-repairing
- A fabric could be programmed to move on its own accord, creating effects such as a flag flapping even without a breeze
- Molecular fasteners could create new clothing design options for truly seamless garments.

This paper summarizes recent advances in molecular manufacturing that are enabling the development of this radical new vision for the future of textile design and manufacture.

BACKGROUND

Although the term "nanotechnology" is now used to describe a broad and diverse range of technological areas, it was originally used to describe a novel method of manufacturing first articulated by Richard Feynman: that molecular machines should be able to build substances by mechanically placing each atom into position exactly as specified [1]. More recently, Eric Drexler has provided a compelling vision of how massively parallel arrays of molecular assemblers could build large, atomically precise objects cheaply and quickly [2-5]. The envisioned products of these molecular manufacturing systems include:

- powerful desktop computers with a billion processors
- abundant energy with inexpensive, efficient solar energy systems

- cures for serious diseases using nanorobots smaller than cells
- new materials 100 times stronger than steel
- a clean environment with nanomachines to scavenge pollutants
- more molecular manufacturing systems (they could build copies of themselves)

Since 1986 this vision of molecular nanotechnology has captured the public's imagination and is now an integral part of popular culture. References to nanomachines are standard fare in many well-known science fiction books, movies, and television shows. Encyclopedias and children's books feature Drexler's colorful and atomically-accurate designs of molecular gears and bearings, as well as artist's renditions of (often fanciful) nanomedical devices cleaning a blocked artery or killing a virus. Today's generation is expecting some form of this vision to happen in their lifetimes.

By contrast, the scientific community has been less than embracing of these ideas. Some scientists claim that this technology is either so distant in the future that we need not concern ourselves, or fundamentally impossible and will never happen. The engineering community has been for the most part silent about the controversy, which is really quite interesting because the proposed technology has very little to do with new science and everything to do with engineering analysis, design, and construction.

My assessment is that although progress is slower than it could be if we had some more focused efforts, there have been significant advances and we will likely see a functioning molecular manufacturing system by the year 2015. In this paper I provide a brief synopsis of the molecular manufacturing vision, show that recent advances and new tools have brought us *past the threshold* of the era of molecular machines, and offer a rather modest vision of what the textile community could do with a desktop manufacturing system that can precisely tailor the molecular structure of a fabric and create *inexpensive* textile products with imbedded:

- nanocomputers and molecular memory
- nanosensors
- micro- and nanomotors and actuators
- solar energy collectors

- wireless receivers and transmitters
- micro and nano-plumbing
- video displays
- energy storage devices

MOLECULAR MANUFACTURING

Molecular manufacturing is a method conceived for the massively parallel processing of individual molecules to fabricate large atomically-exact products. It would rely on the use of many trillions of molecular robotic subsystems working in parallel to process simple chemicals into new materials and devices. Built to atomic specification, the manufactured products would exhibit significantly higher performance than that of today's products. Equally as important, the high level of automation of the manufacturing process would significantly lower the cost compared against today's techniques. A distinguishing feature of molecular manufacturing would be that the trajectory and orientation of every molecule in the system are precisely controlled during the manufacturing operation, differentiating it from processes based on solution chemistry where molecules bump against each other in random orientations until reactions occur.

A few of the key concepts from the principal reference, *Nanosystems*, are summarized in Figures 1-3. Figure 1 shows a cylindrical bearing, a differential gear, and a schematic of a molecular sorting and conveyor transport system. The design and performance of the first two mechanical parts have been studied in detail, and show that high efficiencies are possible when complementary atomic surfaces are properly matched. Figure 2 shows a schematic of a stiff robotic arm composed of about four million atoms. Simple hydrocarbon molecules are fed to its tip through an internal conveyor system; atoms are transferred from those molecules to the workpiece at processing speeds approaching 500,000 atoms/second – about the speed of a fast enzyme. The other diagram shows how assembly stations could be arranged to construct small components at the densely-branched tips and the feed those products into successively larger sub-assemblies. Figure 3 shows a conceptual diagram of a desktop molecular manufacturing system. Simple hydrocarbon molecules are sorted, attached to conveyors, positioned, and then reacted to build up atomically exact structures.

It is particularly appropriate to be discussing this technology with the textile community audience at Ecotextile 04. Of all the manufactured products that come to mind, there is no better analogy to molecular manufacturing than the production of textiles, which assembles tonnage quantities of material from small fibers using up to tens of thousands of machines operating in parallel. It is also relevant to note how clean molecular manufacturing is expected to be – while these systems would manufacture products to atomic specification, they would also prepare waste products to atomic specification. Water vapor and carbon dioxide would be typical waste-stream constituents.

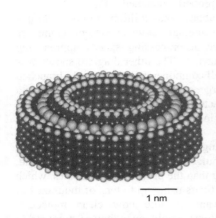

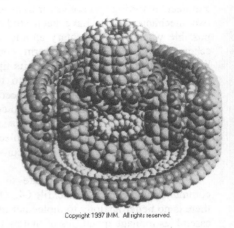

1 nm

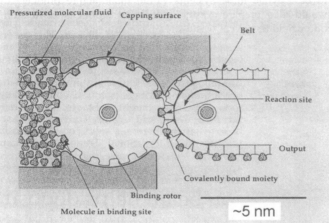

Figure1. Molecular mechanical components of increasing size and complexity. Electromechanical systems will be constructed at the molecular scale, including: support structures, rods, shafts, gears, bearings, conveyors, nanomotors, and manipulators. With proper design and built to atomic specification and precision, sliding surfaces would have low friction and gears and nanomotors would have high power conversion efficiency. The designs shown employ C, H, O, S and N atoms. *Sources: K.E. Drexler, Nanosystems [5], and the Institute for Molecular Manufacturing (www.imm.org).*

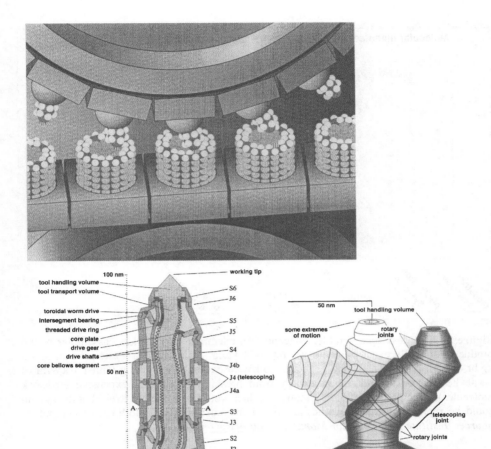

Figure 2. Molecular mills and molecular positioners. Molecular mills (top) would be used for high speed fabrication of highly redundant components. In this illustration a hydrogen atom (striped) is being added to a sleeve bearing as it passes by on a conveyor (which is not drawn to atomic detail). Molecular positioning arms (bottom) would be used to fabricate more customized components. At the nanoscale, megahertz rates of atomic placement are typical and estimates of system performance show that a four million atom manipulator arm (top) could make a copy of itself in less than 10 seconds. This is consistent with the known molecular processing speeds of enzymes in biological systems. *Sources: K. E. Drexler, Nanosystems [5], and Institute for Molecular Manufacturing (www.imm.org).*

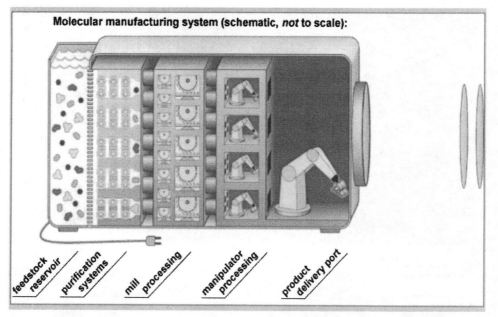

Molecular manufacturing system (schematic, *not* to scale):

feedstock reservoir / purification systems / mill processing / manipulator processing / product delivery port

Figure 3. The desktop assembler concept. An exemplar 1 kg desktop assembler would produce atomically exact products at a rate of 1 kg/hr, would have a waste product of 1.5 kg/hr of high purity water, and generate 3.6 kW/hr excess power along with 1.1 kW/hr of waste heat (from the release of energy from breaking bonds of inexpensive feedstock molecules). One product of the desktop assembler would be a copy of itself, but the system could be reprogrammed to synthesize other items such as food, clothing, or computers. *Source: Institute for Molecular Manufacturing (www.imm.org).*

STATE OF THE ART

In recent years there have been significant theoretical and experimental advances in molecular technologies that are enabling the development molecular manufacturing. Here, we focus on key experimental advances, which are summarized in Tables 1-3. Table 1 provides examples of advances in molecular construction. In 1999 Ho and Lee achieved the first documented instance of positional molecular assembly in a non-biological system, when they picked up a single molecule of carbon monoxide, positioned it over a single atom of iron, and applied a voltage to force a chemical bond to occur between the Fe and CO. Ruoff and Banhart employed electron beams to bind carbon nanotubes to each other and to an atomic force microscope tip, demonstrating a successful joining technique.

Table 2 lists some advances in molecular electronic device development. The 1991 discovery of the carbon nanotube, and subsequent investigations that revealed its novel electronic properties, provided the groundwork for the stunning achievement of the demonstration of a molecular electronic memory based on carbon nanotubes in 2001 by Stan William's group at Hewlett Packard.

Molecular motors are ubiquitous in biological systems (for example, bacterial flagella, or the actin/myosin in our own muscle tissue). Table 3 provides examples showing that we can co-opt a biological motor and attach it to inorganic devices (Montemagno), and create a fully synthetic molecular motor based on a nested carbon nanotube bearing (Zettl). Nested carbon nanotubes can also serve as a telescoping arm; Van der Vaals forces provide a restoring force that cause an extended tube to retreat back to its original position.

Table 1. Experimental Advances in Molecular Construction		
1999	Ho and Lee (Cornell U.) used a scanning tunneling microscope to pick up a single carbon monoxide molecule and chemically bind it to a single iron atom by applying a voltage [7] (see picture, right). This proved the concept of positional assembly using a non-biological robotic system.	
2000	Ruoff's group (Northwestern U.) used an electron beam to attach individual nanotubes to cantilevers, then measured their tensile strength (up to 63 GPa) [8].	
2001	Banhart's group (U. of Ulm, Germany) used an electron beam to attach individual nanotubes to each other [9].	

Research Initiatives in the US. In 1993 Rice University announced the first laboratory in the U.S. dedicated to nanotechnology research, and since then several dozen institutions worldwide have established their own dedicated centers. In 1996 Jim Von Ehr formed Zyvex, the first molecular nanotechnology company. Their goal is to develop the technology and build self-replicating molecular assemblers. Other nanotechnology companies have since been established (or divisions created within larger companies), many focused on (1) molecular electronic devices for computer applications, (2) the synthesis of carbon nanotubes and other fullerenes, and (3) the synthesis of inorganic nanoparticles. In 2000, President Clinton announced the National Nanotechnology Initiative (NNI) with a doubling of funding on nanotechnology research to about $500 million annually. This move cemented the already growing interest in nanotechnology in the United States, and there are similar new initiatives in Europe, Japan, and China. The NNI FY2004 funding level is $847M.

Table 2. Experimental Advances in Molecular Electronic Devices	
1996	The first conductivity measurements of single molecules using an STM [10].
1997	• The first measurement of electronic conduction in a single molecular wire [11]. • The electrical conductivity of carbon nanotubes was demonstrated. • The first molecular diodes were synthesized [12, 13].
1998	Carbon nanotube transistors were made and characterized.
1999	Reversible molecular switches were synthesized and tested (Hewlett Packard/UCLA, Yale/Rice)
2001	Stan Williams' group at Hewlett Packard demonstrated a 64-bit molecular electronic memory [14].
2004	Target year for completion of a DARPA-funded 16 kilobit molecular electronic memory (10^{11} bits/cm^2), now under development [15].

Table 3. Experimental Advances in Molecular Electromechanical Devices	
1999	Carlo Montemagno and George Bachand (Cornell) created the first organic/inorganic integrated molecular motor, using a molecule of the enzyme ATPase coupled to a metallic substrate with a genetically engineered handle. [16]
2001	Alex Zettl's group at Lawrence Berkeley Laboratories developed a nearly frictionless cylindrical molecular bearing (that can also serve as a telescoping arm) based on nested carbon nanotubes [17].
2003	Alex Zettl's group at UC Berkeley developed an electrostatic motor using electron beam lithography to pattern a 100-300 nm gold rotor suspended with a carbon nanotube bearing [18].

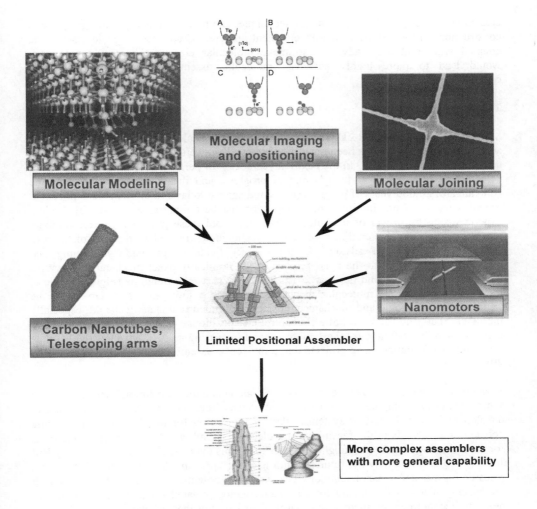

Figure 4. An illustration of the technological advances that are leading to the development of molecular robotic positioning systems.

Figure 4 summarizes recent advances in molecular technologies that have now provided many of the building blocks and construction methods needed to build a positional assembler: structural members, sliding and rotating parts, motors, and positioning and joining technology. Carbon nanotubes, which have been commercially available for several years, can serve as strong, stiff, structural members. They can be joined with electron beams. Individual molecules can be positioned and joined to structures with a scanning tunneling microscope. Nested carbon nanotubes can serve as both cylindrical bearings and telescoping arms. An electrostatic nanomotor has also been created and tested. Molecular modeling techniques can be used to analyze designs prior to construction, further speeding the development process. Crude, less capable assemblers, coupled with continuing advances to make molecular electromechanical components, would lead to more highly advanced molecular assembler systems with broader capabilities.

TIMEFRAME

Molecular manufacturing R&D, and indeed *all* technological progress, is proceeding at an accelerating pace. This has been quantified by Ray Kurzweil, who has shown that the _rate_ doubles every decade [19]. In the next ten years we will perform the equivalent of twenty years worth of research at today's rate of progress, and the next century will see the equivalent of 20,000 years of progress (measured against today's rate). Double exponential growth curves are not intuitive, rendering even the best timeframe guesses by leading experts wrong by orders of magnitude. We can avoid these guesswork errors to some extent by focusing on trendlines for several technologies related to the one of interest. Trendlines for rates of advance in the distinct fields of precision machining and microlithography point to the mass production of atomically exact mechanical structures and computer chips around the year 2015 [20, 21]; Kurzweil's trendline (Figure 5, below) includes the advent of molecular mechanical devices a few years ago. Given (a) the trendlines for electronic and mechanical devices, (b) the current state of the art, especially the fact that the first robotically-controlled positional molecular assembly was demonstrated in 1999, and (c) these more general increasing rates of advance, the Institute for Molecular Manufacturing projects that a molecular assembler could be constructed by 2015.

Barriers to Progress. Despite promises of an economic revolution, and continuing successes in developing molecular robotic components, there are no large scale programs to develop molecular assembler systems. Part of the reason for this has been an ongoing debate in the U.S. about the feasibility of molecular assembly [22, 23], in which the basis for the technology has been brought into question. This may have been exacerbated by Bill Joy's *Wired* article, "Why the future doesn't need us," [24] in which he advocated ceasing research on molecular nanotechnology because of its potential dangers. Threatened with the spectre of losses in funding for all nanotechnology-related research, it was easier to deny the technology completely and marginalize Bill Joy than to clarify the distinctions between various kinds of nanotechnology research and address any dangers directly. European nations seem less concerned about this controversy. For example, in its market analysis Deutsche Bank AG identified molecular manufacturing as one of only three areas of high growth development in nanotechnology [25].

One concern about the ramifications of this debate is that, although development will progress with or without an integrated approach to developing molecular assembler systems, needed safeguards may be omitted without a concomitant focus on proper systems engineering.

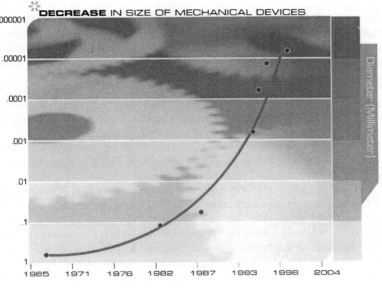

Figure 5. The double exponential decrease in the size of mechanical devices over time (scale is in millimeters). The advent of molecular mechanical devices occurred roughly where the curves intersects 10^{-5}mm (or 10nm--the diameter of a nested carbon nanotube). *Illustration courtesy Ray Kurzweil, KurzweilAI.net.*

IMPLICATIONS FOR TEXTILES

There are clear advantages to having materials that are 100 times stronger than we have now. Objects made from these materials could be up to 100 times lighter, using 100 times less material. As a result, ultralight cars, trucks, trains, and planes would use far less energy, especially with atomically smooth surfaces on moving parts and aerodynamic surfaces to reduce internal friction and air resistance losses.

Textiles will have similar gains in performance. Today, basic units of fabrics are molecules of natural and synthetic materials such as cotton (cellulose), wool (α-keratins), rayon (cellulose), and polyester. Bundles of these molecules are twisted to form fibers, which can be spun into threads and yarns. An obvious way to strengthen these conventional materials would be to reinforce them with carbon nanotubes—the current darling of nanotechnology materials. In fact there are already intensive efforts by groups around the world to create fibers from carbon nanotubes, which individually have a tensile strength of about 100 GPa. This is more than 50 times stronger than a typical steel and 1/3 the density. By comparison, commercial rayon has a tensile strength of 0.45 GPa and nylon, 0.5-0.8 GPa.

One of the difficulties in using carbon nanotubes in textiles is that it is difficult to grow nanotube molecules into centimeter lengths without loss of strength (due to processing-induced defects in long molecules). It's also difficult to twist shorter nanotubes into a fiber while maintaining the high strength of the individual nanotube. With molecular manufacturing, arbitrarily long nanotubes would be possible, and textiles could be fabricated to nearly their theoretical strengths.

Carbon nanotubes also have a high thermal conductivity along the axis (about three times that of diamond, and 15 times that of copper). Like diamond, carbon nanotubes are very stable in air to 1000°C. With these properties, a carbon-nanotube-based textile would make an excellent heat resistant fabric. The high axial thermal conductivity would act as a natural heat pipe to help to dissipate energy from hot spots on the material. Thermal conductivity could be quite low in the transverse plane with an open array of molecules with long, widely spaced cross-links.

Today's textile materials made with molecular manufacturing would be considerably stronger. The theoretical strength of cellulose is 12-19 GPa, so the strength of cotton and rayon could be improved more than ten-fold with molecular manufacturing. As Roger Soderberg has pointed out, there would be virtually 100% efficiency in converting yarns to fabric tensile strength due to the high level of uniformity in both strength and elongation from one yarn to another [26] . Fiber separation could be eliminated as a failure mode by connecting individual fibers end to end and making them continuous, but still bundled and twisted in the same amorphous way. It seems possible to do this while maintaining the look and feel of current fabrics, if desired.

Smart materials and nanotechnology

While synthesis of defect-free materials will lead to substantial improvements in performance, molecular nanotechnology will make more radical changes possible by integrating computers, sensors, and micro- and nanomachines with materials. Here are some ideas:

- Micropumps and flexible microtubes could transport coolant or a heated medium to needed parts of clothing.

- The kinds of sorting rotors shown in Figure 1 could be arrayed as "pores" in a semi-permeable membrane to allow only particular kinds of molecules through. Water might be a useful molecule to select for, to keep one side of a fabric dry or another side wet. On the wet side, the water could be transported away to an evaporator, or stored.

- *Active, programmable materials.* A rich integration of sensors, computers, and actuators within structural materials will blur the distinction between materials and machines, allowing the design and construction of objects that can be programmably reconfigured to sub-micron precision. These materials could monitor and report on their own state of "health." Figure 6 illustrates this concept with a latticework of machines linked by telescoping, interlocking arms. Both information and power would be transmitted through the arms to the individually addressable nodes. By selecting which screws would tighten and which would loosen, the shape of a item could change to conform to the needs of the user. A solid, rigid object could be made to behave like a fabric by effecting rapid changes in its shape, or with temporary disconnections between some cells. A flexible fabric could turn rigid by having loosely bound cells connect into a stiff framework. Thus, distinctions between fabrics and other types of rigid materials could blur.

The programmable material concept is not limited to fabrics but has many potential applications there. One example that Drexler pointed out would be a space suit that would allow nearly as much freedom of movement as one's own skin. Imbedded computers connected to strain gages could sense the wearer's intended movement and adjust the material accordingly. Reflectance of the outer layer could be variable to absorb needed amounts of heat from the sun-facing side and transport it to cold spots—although the material's insulative properties would allow very little of the wearer's heat to escape. Excess heat could be transported to radiators on the cold side.

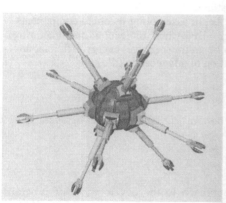 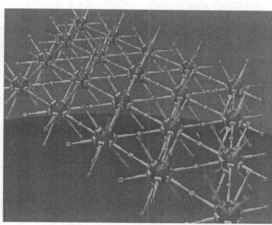

Figure 6. An individual node (left) and a 2D array of interlocked nodes (right). Materials made from these devices could be instructed to change their shape in rapid fashion. *Illustration by J. Storrs Hall, Institute for Molecular Manufacturing.*

- Fabrics could be self-cleaning: robotic devices similar to mites could periodically scour the fabric surfaces and integral conveyors could transport the dirt to a collection site, or the previously mentioned molecule-selective membrane could transport water to one side or the other for a cleaning rinse.

- Fabrics could be self-repairing: sensors would detect discontinuities in the material via loss of signal or a reported strain overload and send robotic "crews" to repair the damage. Self-shaping fabrics would be able to return to their original shape around a tear until repairs are effected.

71

- Large sections of fabrics could be made without visible seams by joining panels of fabric with microscopic mechanical couplings along their edges. Similarly, surfaces could contain mechanical couplings that, when pressed together would bond with nearly the strength of the bulk material. This 'smart velcro' could latch and unlatch at the user's request.

SUMMARY

Molecular manufacturing is an emerging technology that is being developed to build large objects to atomic precision, quickly and cheaply, with virtually no defects. In recent years, molecular manufacturing has progressed from theoretical studies to the construction of working molecular machine components. Positional molecular assembly has been demonstrated in the laboratory, and a potentially useful artificial molecular motor has been synthesized and tested. One company, Zyvex, has announced its intention to design and build a molecular assembler, and many companies are designing and building molecular computer components. Molecular transistors have been built and tested, and a working molecular electronic memory unit has been demonstrated.

Much work remains before molecular robots begin assembling machine components, ushering in the era of molecular manufacturing. But when molecular manufacturing matures, probably by the year 2015, it promises to be an inexpensive, energy efficient, and environmentally benign way to make a new generation of advanced textiles.

REFERENCES

1 R P Feynman, 'There's plenty of room at the bottom', *Engineering and Science*, 1960 **23** 22-36.

2 http://www.e-drexler.com/

3 K E Drexler, 'Molecular engineering: An approach to the development of general capabilities for molecular manipulation', *Proc. Natl. Acad. Sci. USA,* 1981 **78**(9) 5275-5278. (http://www.imm.org/PNAS.html)

4 K E Drexler, 'Engines of Creation', Doubleday, Garden City, NY (1986). (http://www.foresight.org/EOC/)

5 K E Drexler, *Nanosystems: Molecular Machinery, Manufacturing, and Computation*, John Wiley & Sons, Inc.: New York (1992). See http://www.foresight.org/Nanosystems/

6 D R Forrest, 'Molecular Machines for Materials Processing', *Advanced Materials & Processes*, 1993 **141** (1)

7 Ho and H J Lee, 'Single-bond formation and characterization with a scanning tunneling microscope', *Science*, 1999 **286** 1719-1722.

8 M F Yu, O Lourie, M J Dyer, K Moloni, T F Kelly, and R S Ruoff, *Strength and breaking mechanism of multiwalled carbon nanotubes under tensile load, Science*, 2000 **287** 637-640.

9 T Smalley-Brown, 'Electron Beam Welds Nano-tubes', *Technology Research News*, August 1/8, 2001.
http://www.trnmag.com/Stories/080101/Electron_beam_welds_nanotubes_080101.html

10 L A Bumm, J J Arnold, M T Cygan, T D Dunbar, T P Burgin, L Jones II, D L Allara, J M Tour, and P S Weiss, 'Are single molecular wires conducting?' *Science*, 1996 **271** 1705-1706.

11 M A Reed, C Zhou, C J Muller, T P Burgin, and J M Tour, *Science*, 1997 **278** 252.

12 R M Metzger, B Chen, U Höpfner, M V Lakshmikantham, D Vuillaume, T Kawai, X Wu, H Tachibana, T V Hughes, H Sakurai, J W Baldwin, C Hosch, M P Cava, L Brehmer, and G J Ashwell, 'Unimolecular electrical rectification in hexadecyl-quinolinium tricyanoquinodimethanide', *J. Amer. Chem. So.*, 1997 **119** 10455-10466.

13 C Zhou, M R Deshpande, M A Reed, L Jones II, and J M Tour, 'Nanoscale metal/self-assembled monolayer/metal heterostructures', *Appl. Phys. Lett*, 1997 **71** 611.

14 D Berman, 'Beyond silicon: hp researchers announce breakthroughs in molecular electronics', Sept. 2002.
http://www.hp.com/hpinfo/newsroom/feature_stories/2002/molecular02.html

15 http://www.darpa.mil/mto/solicitations/BAA03-12/S/Section1.html

16 C D Montemagno, and G D Bachand, 'Constructing nanomechanical devices powered by biomolecular motors', *Nanotechnology*, 1999 **10** 225-331.

17 http://www.lbl.gov/Science-Articles/Research-Review/Magazine/2001/Fall/features/02Nanotubes.html

18 A M Fennimore, T D Yuzvinsky, Wei-Qiang Han, M S Fuhrer, J Cumings, and A Zettl, Rotational actuators based on carbon nanotubes', *Nature* **424** (July 24): 408-410. See also
http://www.berkeley.edu/news/media/releases/2003/07/23_motor.shtml
http://www.berkeley.edu/news/media/releases/2003/07/video/nano_bband.mov

19 http://www.kurzweilai.net/meme/frame.html?main=/articles/art0134.html

20 R W Keyes, 'Miniaturization of electronics and its limits', *IBM J of Research and Development*, 1988 **32** (1) 24-28.

21 N Taniguchi, 'Future Trends of Nanotechnology', *International J of the Japan Society for Precision Engineering*, **26** 1992 (1) 1-7.

22 A Debate About Assemblers (2001). http://www.imm.org/SciAmDebate2/index.html

23 Rudy Baum, 'Nanotechnology: Drexler and Smalley make the case for and against 'molecular assemblers,'' *Chemical & Engineering News*, 2003 **81** (48) 37-42. http://pubs.acs.org/cen/coverstory/8148/8148counterpoint.html

24 D R Forrest, 'Perspectives on Nano2002:the Sixth International Conference on Nanostructured Materials', 16-21 June 2002. http://www.nanoindustries.com/feature/Nano2002.html

25   Joy, Bill, 'Why the future doesn't need us', *Wired*, April 2000 8.04, http://www.wired.com/wired/archive/8.04/joy.html

26 Personal correspondence with Soderberg, R., 18 August 1995.

THE SUBSTITUTION OF HEMP AND FLAX FOR COTTON IN WOUND SPOOL FILTERS

C. Sevajee* and R. Edyvean
Department of Chemical and Process Engineering, University of Sheffield, Mappin Street, Sheffield, S1 3JD, England

INTRODUCTION

Cartridge filters are used in applications where the prime concern is to maintain a clean fluid rather than to recover solids[1]. They feature high efficiency but relatively low capacity to hold solids[2]. The objective of this paper is to investigate the possibility of producing cartridge filters developed using natural hemp and flax fibre and comparing the performance of prototype cartridges to the filtration performance of cartridge filters constructed from cotton. This has lead to the development of new cartridge filters constructed from hemp and flax fibres for the solid-liquid fluid clarification market. The aim would be to replace imported cotton with UK hemp and flax fibres as the raw material for the manufacture of cartridge filters. The fibres produced by these plants have been chosen because they are vegetable fibres that are cellulosic in composition, have properties that are similar to cotton making them an attractive substitute and they require less intensive farming methods rendering them more environmentally sustainable.

RAW MATERIAL

Cotton

Cotton has been the most popular fibre since the beginning of times. No other fibre comes close to duplicating all of the desirable characteristics combined in cotton. Physically the individual cotton fibre consists of a single long cell, with one end attached directly to the surface of the seed. From the field, the seed cotton is sent for the separation of lint and seed. The cotton first goes through dryers to reduce moisture content and then through cleaning equipment to remove foreign matter. The cotton is then air conveyed to gin stands where revolving circular saws pull the lint through closely spaced ribs that prevent the seed from passing through. The lint is removed from the saw teeth by air blasts or rotating brushes, and then compressed into bales. Lint from several bales is mixed and blended together to provide a uniform blend of fibre properties.

Hemp and Flax

Bast fibres possess by nature a high degree of variability. These fibres are constructed of long thick-walled cells which overlap one another; they are cemented together by non-cellulosic materials to form continuous strands that may run the entire length of plant stem. The strands of bast fibres are normally released from the cellular and woody tissue of the stem by a process of natural decomposition called retting (controlled rotting)[3]. This is the most important process in the preparation of the fibre. It consists essentially in submitting the straw to the action of water and allowing fermentation to take place. The connection between the bast fibres and the woody core is loosened by micro organisms after adding appropriate amount of water. As a result, the bast fibre is separated from the core of the plant, the connection between different parts of the plant is loosened. Once the retting process is complete the retted stalks are dried before it is subjected to the breaking process where the aim is to separate as much as possible the flexible fibre part and the solid core portion of the plant in a sequence of squeeze, break and scutching processes. By using rollers, the decorticated fibres are made softer.

Two traditional methods used to ret hemp and flax are water and dew retting. In water retting, the stems are submerged in rivers and lakes, and anaerobic bacteria colonize the stems and degrade pectins and other matrix compounds, freeing fibres from the core tissues. Dew retting depends on the removal of matrix materials from the cellulosic fibres before cellulolysis, which causes the weaking of the fibres. Water retting results in high quality fibre but has been discontinued in western countries because of the extensive stench and pollution from the fermentation products. Dew retting produces coarser and lower quality fibres with poor consistency in fibre characteristics than water retting[4]. Because of problems with both water and dew retting, extensive research is being carried out to develop enzymatic and mechanical retting to produce fine staple equivalent to cotton.

YARN FORMATION

Spinning converts a mass of short fibres into yarns which requires a number of stages. Initially the staple fibres undergo carding in which they are disentangled, straightened into length-wise alignment and any impurities are removed to produce card sliver. After carding, several slivers are combined. The sliver is then put through an additional straightening called combing. The combing process forms a comb of sliver which in turn produces a smoother and more even yarn before undergoing the drawing process. Drawing produces long thin slivers which are twisted and wound onto bobbins. After drawing, the fibres are spun using a spinning system specific to the end use of the yarn[5]. There is an extensive range of different spinning systems. Important aspects of any spinning system are the fibre types that can be spun, the count range, the economics of the process and the suitability of the resulting yarn structure to a wide range of end users. In the spinning process, there is always a fixed relation between the weight of the original quantity of fibre and the length of yarn produced from that amount of raw material. This relation indicates the thickness of the yarn.

The cotton yarn used is spun using the open-end spinning system, more specifically friction spinning where the free end of the yarn is rotated while individual fibres are collected and twisted onto the end to increase the yarn length. The fibre deposition does not result in a straight and parallel arrangement of the fibres in the spun yarn. As a result it is only suitable for very open coarse count yarn[6]. The conventional ring spinning technique is used to form the hemp and flax yarns, the main reason being the suitability of this technique to produce the required yarn structure. The ring spinning technique is effectively a twisting and winding mechanism where the ring spinning frame completes the manufacture of yarn by drawing out the roving, inserting twist and by winding the yarn onto a large bobbin forming what is referred to as a cheese.

NATURAL FIBRES

When considering the use of natural fibres for filtration purposes the important factors to consider are the fibre properties, both mechanical and chemical properties, fibre dimensions, fibre density and the manufactured yarn properties. Desirable properties are that the fibre length should not be shorter than 6-12 mm and that the fibre length should be several hundred times the width to give a high L/D ratio. These two conditions are important. The absence of either would not allow the fibre to hold together which would not allow fibres to be twisted together to form a yarn[3]. The fineness of the fibre, which is dependent on the width of the fibre, enables the determination of how fine a yarn can be produced. Table 1 gives the dimensions of the three chosen fibres. There is a large variation in fibre dimension as it is an inherited characteristic that can be greatly influenced by soil and weather[7].

Table 1. Dimensions of fibres[7,8]

| Dimensions | Length, | Width, mm | Fineness, μm | | L/D |
Fibres	Range	Average	Minimum	Maximum	Ratio
Cotton	1.5-5.6	0.012-0.025	2.82	8.46	2500
Hemp	100-300	0.06-9.04	8.46	56.4	1078
Flax	20-140	0.04-0.62	5.64	45.1	1778

CHEMICAL PROPERTIES

Natural fibres of vegetable origin are constituted of cellulose with varying amounts of other natural substances such as lignin, pectin, hemicellulose, waxes and gums[8]. The amount of these associated substances and the ease with which the cellulose fibre can be separated from them determine how useful the fibre can be[3]. Hemp and flax are lignocellulosic fibres where the cell wall is made up mainly cellulose, hemicellulose and lignin. Cotton is a non-lignocellulosic fibre that does not contain lignin. Cellulose is the basic structural component of plant fibres and has a molecular weight $> 10^6$ g/mole[9]. It is the most important organic compound produced by plants. The cellulose molecules consists of glucose units linked together in long chains, which in turn are linked together in bundles called microfibrils. The structure of cellulose is shown in Figure 1[9]. Cellulose molecules run parallel to one another and forms spiral that are highly parallel to one another in vegetable fibres. Flax and hemp contain a lower percent of cellulose than cotton and the spiral angle of the cellulose is lower than that of cotton. A lower helix angle increases the stiffness of the fibre, accounts for low extensibility and makes the fibre brittle[7].

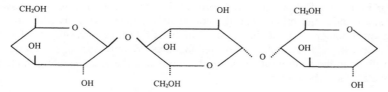

Figure 1. Chemical structure of cellulose

Hemicelluloses are also found in all plant fibres. Hemicelluloses are polysaccharides bonded together in relatively short, branching chains. They are intimately associated with the cellulose microfibrils, embedding the cellulose in a matrix. Hemicelluloses are very hydrophilic. Lignin, the Latin word for wood, is the compound which gives rigidity to the plant[10]. The chemical composition of the three fibres are given in Table 2[7].

Table 2. Chemical composition of fibres[11,12]

Fibres	Cellulose	Hemi-cellulose	Pectin	Lignin	Extractives
Cotton	91.8	6.4		0	1.8
Hemp	74.4	17.9	0.9	3.7	3.1
Flax	71.2	18.6	2.0	2.2	6.0

MECHANICAL PROPERTIES

Yarns are built up by twisting together the long, thin flexible but strong things called fibres therefore the properties of the yarn depends very largely on the properties of the fibres from which it is made[3]. Strength is needed to enable the fibres to withstand the spinning process and to provide strength in the final yarn. Flexibility permits the fibres to be spun. Flax is a stronger fibre than cotton. It is a particularly inextensible fibre and it stretches only slightly as tension increases. It has a high degree of rigidity and resists bending. Hemp is a coarser fibre than flax. The fibre is strong and durable more lignified than flax and is consequently stiffer. Certain mechanical properties of the three fibres are shown in Table 3.

Table 3. Mechanical properties of fibres[3,9,11,12]

Fibres	Density (g/m^3)	Ultimate strain at break (%)	Tensile strength (g/dtex)	Moisture adsorption (%)	Spirality	Spiral angle (^{0})
Cotton	1.52	5-10	3-6	8	Z	20-30
Hemp	1.48	2-4	5-6	8	Z	5-10
Flax	1.51	2-7	5-6	7	S	5-10

STRUCTURE AND APPEARANCE

The surface structure of a fibre is most important in that it controls the behaviour of the fibres in the yarn. The cross-sectional shape of a fibre has an important influence on its behaviour in a yarn. Cotton has a flat, ribbon-like cross section. Figure 2 shows that when viewed under the microscope the cotton fibre appears to be twisted frequently along its entire length with the direction of the twist reversing occasionally. Twists are referred to as convolutions and it is believed that they are important in spinning because they contribute to the natural interlocking of the fibres in the yarn, enabling the fibres to grip to one another when spun[7].

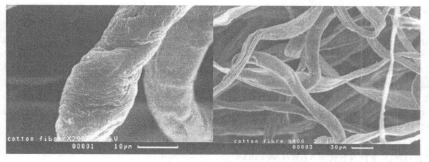

Figure 2. Cotton fibre SEM

In Figure 3 the hemp fibres are seen to be irregular in shape, being flattened at some points along its length but cylindrical at others with joints, crack, swelling and other irregularities on the surface[3]. There are striations on the surface of the fibre but no nodes like those found in flax[13].

Flax is polygonal in cross-section. Seen under the microscope in Figure 4, the fibre cells show up as long cylindrical tubes which may be smooth or striated lengthwise. They do not have convolutions which are characteristic of cotton. The width of the fibre may vary several times along its length. There are swellings or 'nodes' at many points and the fibres show characteristic cross-markings[3].

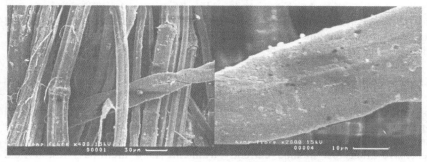

Figure 3. Hemp fibre SEM

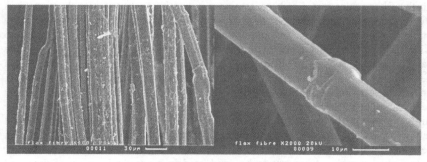

Figure 4. Flax fibre SEM

ENVIRONMENTAL BENEFITS

The farming of cotton requires the use of large volumes of agrochemicals to provide protection against pests and also large volumes of irrigation water to maximise fibre yields. Both these requirements have damaging effects on the environment. Hemp and flax have considerably less requirements for intensive agricultural intervention and almost zero requirements for fertilisers and pesticides. Hemp and flax are also useful break crops being beneficial in crop rotation and are sustainable. And the biomass production of these plants recycles CO_2, cleaning the atmosphere.

CARTRIDGE FILTER FABRICATION

Spool-wound cartridge filters are constructed from a length of yarn that is wound around a central perforated core. The yarn consists of intertwined fibres of 15-20 μm in diameter, the fibre diameter being important in determining the particle retention efficiency of the final product. This yarn is teased to produce a nap and wound around the core in a precisely controlled winding pattern. A diamond pattern is the most common configuration used. This creates a graded pore structure which decreases in size from outer, upstream surface to the inner central core. Proper brushing of the yarn provides a high surface area for adsorptive filtration as well as tortuous flow channels for particle interception. The performance of a filter depends upon the type of fibre and the way the yarn is produced and wound[2]. In addition factors such as fibre diameter and length, yarn diameter, tension and winding technique all effect the flow path, shape and size, as well as the amount and nature of internal surfaces available for particle collection. Such characteristics can be varied throughout the depth of the filter medium.

DEVELOPMENT IN STANDARDISED FILTER TESTING

The test rig designed for this work is shown in Figure 5. The test system is a modified single-pass version of the test system described in ISO 4572 (1981)[14]. ISO 4572 describes the multipass test system that has become accepted as the most reliable and comprehensive filter test method currently available. In a multi-pass test a fluid containing a known quantity of particulate contaminant is continuously circulated through the filter under test. Fresh contaminant is continuously fed into the system in order to replace what has been removed from the fluid by the filter. Samples of the influent and effluent streams are continuously collected and analysed in order to determine the number and size of particles in the test fluids, from which the particle removal efficiency (%) of the filter may be calculated. In addition, this test also provides information on filter element life, dirt holding capacity and pressure tolerance. The modified test system utilises a very high efficiency clean-up filter that removes any particles remaining in the process fluid after passage through the test filter. This mode of testing represents the most frequent conditions of use of cartridge filters in real applications[9]. A further modification to the test system is the use of alternate upstream particle concentrations of 25 mg/l for 30 minutes and 3 mg/l for 30 minutes. This test, enables a single cartridge filter test to be conducted in a practical period of time and provides useful data on the performance of the filter during periods of both high and low upstream particle concentrations.

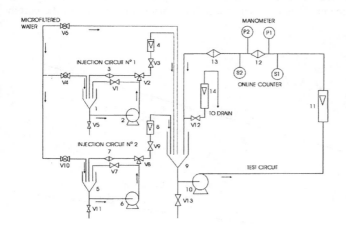

Figure 5. Design of the filter test system showing the injection circuit reservoirs (1,5), dosing pump (2,6), Clean up filters (3,7,13), dosing flow meter (4,8), test circuit reservoir (9), metering pump (10), test flow meter (11), test filter (12), flow meter to drain (14), pressure ports (P1, P2), sampling probes (S1, S2), valves (V1-V13)

The filters are tested by using pre-filtered water as the test fluid at a constant temperature of 25^{0}C and a flowrate of 16 l/min. A standardised test dust (ISO Medium Test Dust) with a particle size range between 0 to 120 μm is used at a upstream concentration of 3 mg/l. This particle concentration is maintained for 30 minutes after which point the contaminant slurry is fed from a separate reservoir providing an upstream concentration of 25 mg/l for 30 minutes. After this time the feed slurry is switched back to the first feed reservoir providing a particle concentration of 3 mg/l for 30 minutes and so on. This practice ensures that the filter will attain a terminal differential pressure at 2 bar at the completion of the test. The filtration efficiency of the filter is monitored throughout the test by measuring the number and size of particles in the fluid downstream of the filter throughout the test with a Liquilaz automatic particle size analyser (Particle Measuring Technique (GB) Ltd.) The same measurements are taken for the fluid upstream of the filter when the particle concentration in the upstream fluid is 3 mg/l, since the higher particle concentration will exceed the sensitivity of the particle size analysis equipment which is 15,000 particles/ml[10].

Standard filter performance tests were carried out using wound cartridge filters constructed from hemp and flax yarn and compared with filters produced from cotton. Cotton cartridge filters nominally rated at 1, 5, 10 and 20 micron were constructed using 1000 tex yarn. Hemp and flax filters were then made up by the same manufacturer based on the standard winding pattern used for cotton cartridge filters. For the purpose of this research several modifications were carried out. Firstly due to the natural physical properties of the bast fibers, i.e the fiber length and fiber diameter, the yarn was constructed using the conventional ring spinning system because of its ability to process most natural fibers for the majority of end uses.

COMPARISON OF SPOOL WOUND CARTRIDGE FILTERS

For this paper, identical cotton, hemp and flax filters constructed from 1000 tex yarn and nominally rated at 5 micron will be compared and contrasted. The first batch of hemp yarn produced was made to match the identical cotton yarn. Upon testing the filters constructed and after further investigation was carried out, it was discovered that where the cotton yarn was made up to have 80 turns per meter the hemp yarn was twisted 116 times per meter. This resulted in a tighter yarn configuration to the cotton yarn.

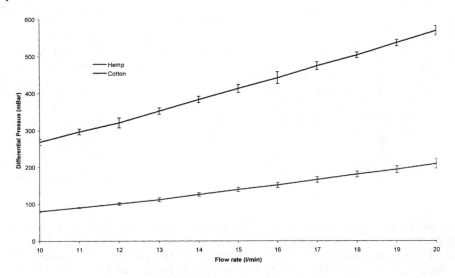

Figure 6. Clean element differential pressure of "5 micron" cotton (top line) and hemp filters (bottom line). Cotton yarn=80 turns/meter, hemp yarn=116 turns/meter.

The cotton and hemp clean element differential pressure is displayed in Figure 6. This differential pressure is also known as the initial head loss and corresponds to the resistance to fluid flow through the clean filter medium which is directly proportional to fluid velocity[15]. The cotton cartridge filter displays a considerable high clean element differential pressure which will result in a shorter element life when compared to the hemp filter. The higher clean pressure drop of the cotton filter is due to the nature of spinning system used to form the yarn. Friction spinning process results in a bulky yarn where the fibres are individually collected and twisted onto the yarn when compared to the ring spun yarn where the yarn is a collection of straight and parallel fibres twisted together which results in a neat yarn. Though the hemp filters were made up using the same production technique as the identical cotton filters, it is seen that the filters do not behave in a similar manner. As a result it will not be possible to give the same rating to the hemp filters as one would using the winding specifications used to construct the identical cotton filter.

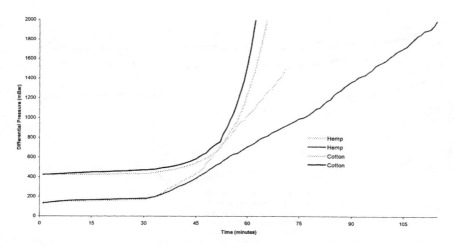

Figure 7. The element life of "5 micron" cotton (top solid and dashed curves) and hemp filters (bottom dashed and solid curves). Terminal pressure=2000 mBar, cotton yarn=80 turns/meter, hemp yarn=116 turns/meter.

The duration of the filter test in minutes is the time taken for the pressure differential across the test element to attain the terminal pressure differential. In this case the terminal pressure differential is defined at 2000 mBar. The change in pressure differential throughout element life as solids accumulated both on and within the filter structure is represented graphically in Figure 7. The cotton curve represents the classic life curve of a spool wound cartridge with little accumulation in the pressure differential throughout the majority of life until approximately 60% of the net pressure differential. At this point the increase in pressure differential in time is very rapid until attainment of the terminal pressure differential. The capacity of the filter is mostly consumed before the sharp increase in pressure differential. Hemp filters show an increase in pressure differently gradually. After the low concentration cycle for the first 30 minutes of its life the hemp cartridge displays a consistent accumulation in pressure differential throughout the entire duration of its life which results in the hemp cartridge having a longer element life than the cotton cartridge. A lower initial clean pressure drop displayed by the hemp cartridge in Figure 6 is a contributing factor to the longer element life until it reaches terminal differential pressure. The maximum allowable pressure tolerance of the filter is the limit beyond which the filter will fail structurally and must not be exceeded.

Analysis of the test fluid from upstream and downstream of the test element by the Liquilaz automatic particle size analyser allows the calculation of particle retention efficiency of the element which is displayed graphically in Figure 8 which demonstrates the variation in the retention of particles of different sizes. The particle retention efficiency is found to increase for larger sized particles. The element appears to become more efficient at retaining larger sized particles.

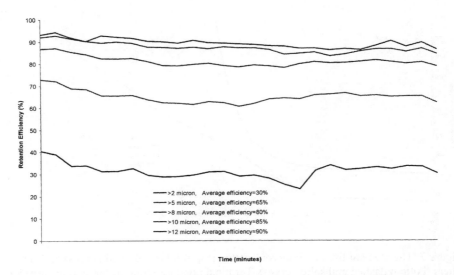

Figure 8. Retention efficiency of "5 micron" hemp filters (results shown in descending order, starting with >2 micron). Hemp yarn=116 turns/meter.

Upon comparison of the particle retention efficiencies attained by both the cotton and hemp filters tested it was seen that where the cotton filters displayed a retention efficiency of greater than 99% for all particles greater than 5 micron in size the identical hemp filter produced an efficiency of 65%. As these were the first production batch of the hemp filters, alterations to the yarn production and winding specifications is hoped to improve the performance even more. And as it was decided to initially concentrate on the hemp yarn and then use the information learnt to develop the flax yarn, latter modifications were applied in the production of the flax yarn as well. Though the cotton cartridges were wound using yarn having 1000 tex count, hemp and flax yarn were made to three different finesses, 1000, 800 and 500 tex. Ultimately this will allow initial comparison testing of cotton, hemp and flax cartridge filters of the same micron rating having the same yarn count which will provide a basis for changes in yarn count. The amount of twist applied to the production of the new hemp and flax yarn was made to match the cotton yarn having 80 turns per meter. Hemp and flax cartridges constructed from the modified yarn was tested and the results obtained analysed.

The clean element differential pressure of identical 5 micron cotton, hemp and flax filters constructed from 1000 tex yarn is displayed in Figure 9. Once again the cotton cartridges display a high clean element differential pressure while both the hemp and flax cartridges displayed a much lower differential pressure. Upon further investigation it was seen that the hemp cartridges initially constructed from the 1000 tex yarn with 116 turns per meter displayed a higher clean differential pressure to the cartridges constructed from the yarn modified to have a lower number of twist.

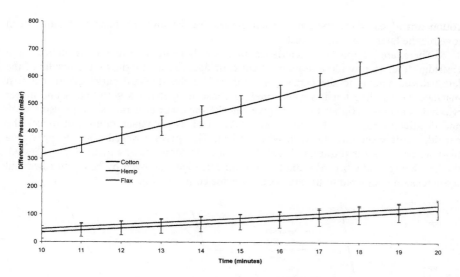

Figure 9. Clean element differential pressure of "5 micron" cotton (top line), hemp (middle line) and flax (bottom line) filters

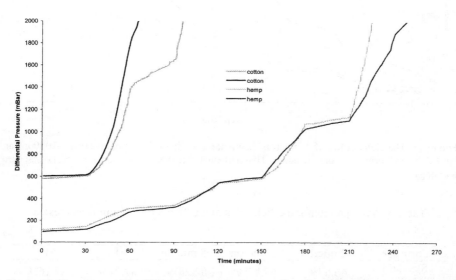

Figure 10. The element life of "5 micron" cotton (top two lines) and hemp (bottom two lines) filters (Terminal pressure=2000 Bar)

Figure 10 shows the increase in element life of the filters tested as particles accumulate in the filters pores. This figure enables us to visualise the speed of clogging which is a major element in the comparison of three cartridges when subjected to the same mass flow of solid particles. As shown previously the hemp filter elements exhibits a more continual accumulation in differential pressure when compared with cotton cartridges where the cotton filter curves are flat initially, becoming more steep towards the later stages of the test.

There are various characteristics that can be varied in the construction of a cartridge filter to provide a more efficient cartridge. Figure 11 shows the results of the latest developments to the cartridges compared to the standard cartridge. In the first instance the winding tension was varied, to construct cartridges with higher and looser winding tensions. Higher tensions give finer ratings, but also raises the pressure drop and drastically reduce filter life. Loose tension gives a coarser rating but makes the cartridge soft. Secondly the yarn was brushed. This gives a raised effect to provide a high surface area for filtration. Three different configurations of cartridges were tested, the first being heavy brushed and top tension cartridge, followed by heavy brushed and loose tension compared to the standard cartridge configuration.

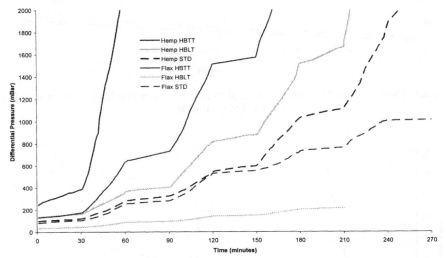

Figure 11. The element life of "5 micron" hemp and flax filters (Terminal pressure=2000 mBar, HBTT=heavy brushed, top tension, HBLT=heavy brushed, low tension, STD=standard winding)

Table 4. Average retention efficiencies of the "5 micron" hemp filters tested

Cartridge	>2 micron	>5 micron	>8 micron	>10	>12
Standard winding	46.67%	69.82%	82.28%	87.21%	87.16%
Heavy brushed, low tension	45.52%	68.59%	84.89%	90.28%	94.04%
Heavy brushed top tension	57.49%	78.21%	93.16%	97.86%	99.16%

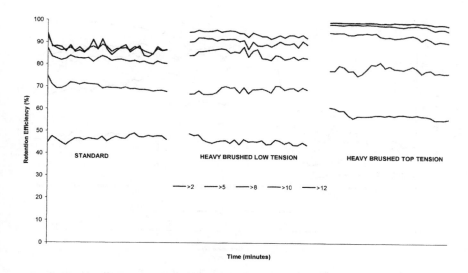

Figure 12. Retention efficiency of "5 micron" hemp filters of various modifications
(results shown in descending order, starting with >2 micron)

The particle retention efficiencies of the various 5 micron hemp filters tested are displayed graphically in Figure 12 and the average retention efficiencies are given in Table 4. The heavy brushed cartridge with the higher tension displayed the highest capture efficiencies for all particle sizes while the heavy brushed loose tension and standard wound cartridges displayed similar capture efficiencies for particles sizes up to 8 micron after which the standard wound cartridge managed to retain the same percentage of particulate for particles greater than 10 micron while the particle removal efficiency displayed by heavy brushed, loose tension filter was found to increase for larger sized particles.

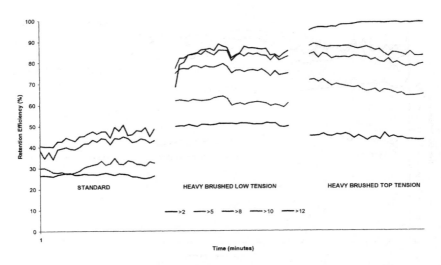

Figure 13. Retention efficiency of "5 micron" flax filters of various modifications (results shown in descending order, starting with >2 micron)

Table 5. Average retention efficiencies of the "5 micron" flax filters tested

Cartridge	>2 micron	>5 micron	>8 micron	>10	>12
Standard winding	26.66%	30.78%	41.24%	45.37%	ND
Heavy brushed, low tension	50.58%	61.39%	76.40%	83.11%	84.20%
Heavy brushed top tension	44.55%	67.38%	81.12%	85.83%	97.86%

The efficiencies of the various flax filters tested are displayed in Figure 13 and Table 5. Once again it is seen that the heavy brushed, tight tension cartridge displays the highest particulate capture efficiencies, followed by the heavy brushed, loose tension cartridge and finally the standard wound cartridge. Overall it is seen that the flax filters exhibit a lower retention efficiency that the identical hemp cartridge filters for all the particle sizes analysed.

The improved performance found when cotton filters were tested was attributed to the nature of the cotton fibres. The striated surface of the cotton fibres[7] and their ability to swell on interaction with water[2] ensures that the filter structure becomes tighter and less porous during filtration. The surface structure also facilitates the collection of particulate material in the ruffles along the surfaces of the fibres. This probably contributes to the accelerated accumulation in differential pressure even early in the life of cotton filters compared to hemp elements. Though the hemp fibre also shows characteristics of striations on its surface there are many other irregularities exhibited which include joints, cracks and swelling along its length which could present problems in particles attaching to its surface. The occurrence of dislocation or joints throughout the length of the flax fibre and its smooth surface when compared to the hemp fibre could be the reason for the reduced particulate capture efficiency displayed by the flax cartridges.

CONCLUSION

It can be said that the substitution of hemp and flax for cotton in spool wound cartridge filters has generated encouraging results. But in order to produce a competitive alternative to cotton and to use the multiple properties of the hemp and flax fibres the development or the application of processing methods is indispensable. Preparation and processing characteristics must be greatly improved to match bast fibres to those of cotton. The bast fibres and yarns used to construct the cartridges do not fulfil the requirements of the filtration industry. In regards to these wide distribution of fibre diameter and length the filtration industry is faced with vast processing problems. To enable an application of bast fibres, the advantages of the cotton fibre must be taken into account. For example, raw bast fibres can be shortened using a fibre cutting machine and sticky substances such as pectin and lignin must be removed and the fibres must be cottonized to gain almost complete isolation of single fibres such as those of cotton fibre[16].

REFERENCES

1. E Fochtman, 'Cartridge filtration: Technology and Economics', *Filtration and Separation,*, 1973 **10**(2) 289-294.

2. A C Shucosky, 'Select the right cartridge filter', *Chemical Engineering,* 1988 **95**(1) 72-77.

3. J Gordon-Cook, *Handbook of Textile Fibres*. Durham, Merrow, 1984.

4. C Van Sumere, *The biology and processing of flax*, Belfast, M Publishing, 1992.

5. B P Corbman, *Textiles Fibre to Fabric*. New York, McGraw Hill Book Company, 1983.

6. C A Lawrence, *Fundamentals of Spun Yarn Technology*, Boca Raton, CRC Press, 2003.

7. B A Kottes and I V Gruy, *Encyclopedia of Textiles, Fibres and Nonwoven Fabrics*, New York, John-Wiley & Sons, 1984.

8. J McGovern, *Encyclopedia of Textiles, Fibres and Nonwoven Fabrics*, New York, John-Wiley & Sons, 1984.

9. S B Warner, *Fibre Science*, New Jersey, Prentice Hall. 1995.

10. P O Olesen and D V Plackett, 'Perspectives on the performance of Natural plant fibres', *Natural Fibres Performance Forum Plant Fibre Products - Essential for the Future,* Copenhagen, Denmark, 1999.

11. E H Lloyd and D Seber, 'Bast fibre application for composites', *Proceedings of the 30th Particle Board/Composite Material Symposium,* WSU, Pullman, WA 1996.

12. N Kerr, 'Evaluating textile properties of alberta hemp', *Canadian Textile J,* 1999 **116**(1) 36-38.

13. J Militky, V Bajzik and D Kremenakova, 'Selected properties of cottonized flax', Proc. *International Seminar Netecoflax,* Covilha, Portugal, 2002.

14. *Hydraulic fluid power- Filters - Multi-pass method for evaluating filter performance,* International Standards Organisation, ISO 4572,1981.

15. K J Ives, 'Deep bed filtration: principles and practices', *Filtration and Separation,* 1980 **17**(2) 157-168.

16. K M Nebel, 'New processing strategies for hemp', *J of the International Hemp Association,* 1995 **2**(1) 1, 6-9.

IT MAY BE ECO-FRIENDLY BUT IS AN INGEO ™/CELLULOSE BLEND STRONG ENOUGH TO WITHSTAND WET PROCESSING?

Jantip Suesat and Duncan Phillips
Department of Textiles and Paper, University of Manchester, P O Box 88, Manchester M60 1QD, UK

ABSTRACT

The development of environmentally-sustainable poly (lactic acid) fibres, enable them to compete with conventional polyethylene terephthalate fibers. However, PLA fibers are more sensitive to wet processing treatments that PET fibers and this paper discusses how successful dyeing of this former may be achieved.

INTRODUCTION

Poly(lactic acid) (PLA) is a synthetic polyester derived from renewable sources such as corn. PLA has been used for many years in biomedical applications such as drug release systems, medical implants and surgical sutures because of its biodegradability and biocompatibility. It can be recycled either by re-melting or by hydrolysing into the starting monomer, lactic acid. Nowadays, apart from being used in medical applications, the versatility of PLA is exemplified by its use in packaging outlets as well as in a series of textile and agricultural applications [1,2]. As a consequence of its interesting properties and its potentially "green" credentials, PLA is being increasingly considered for use as a commodity polymer.

PLA fibres are expected to compete with petroleum-based materials both on a cost and performance basis, representing the first material to successfully bridge the gap between natural and synthetic fibres in its technical properties. Consequently, it has the potential to provide the textile industry with a novel product option, viz. an environmentally-friendly synthetic (polyester) polymer, which can be used in a range of apparel and non-apparel fabric applications [3,4]. Many of its properties are comparable to those of polyethylene terephthalate (PET) fibre e.g. its dimensional stability and crease resistance[3]. However, because it possesses superior moisture management properties, fabrics containing PLA are being considered for use in sportswear as well as other outerwear outlets. PLA fibres are marketed by Cargill Dow under the brand name INGEO and the work carried out at the University of Manchester has been done on samples of both fibres and fabrics supplied by Cargill Dow. Blends of PLA with both cotton and wool are also being investigated in order to produce higher value-added fabrics, which exhibit the combined characteristics of the individual fibres (dimensional stability and comfort).

Compared with PET, PLA is a 'greener' polymer, being produced from plants and food crops. After manufacture and subsequent processing of the polymer into the desired products, it can later be degraded into soil and humus, hence encouraging the re-cycling process. It does not suffer the bio-accumulation problems associated with the waste derived from conventional polyester, PET, which, of course, is derived from finite petroleum resources.

PROPERTIES OF PLA

The properties of PLA, as a textile fibre, in comparison with those of conventional polyester (PET) are shown in Table 1. PLA has better elastic recovery, better hydrophilicity, higher limiting oxygen index (LOI) and lower smoke generation than PET. Its lower refractive index allows PLA often allows it to be dyed to a deeper shade than PET, using a given concentration of disperse dye on fibres of similar dimensions. However, the melting temperature (T_m) of PLA, being much lower than that of PET (Figure 1), means the need to use much lower domestic ironing temperatures, particularly on woven constructions. Another disadvantage of PLA is its higher sensitivity to alkali, as compared with PET, resulting in some loss of fibre strength during subsequent wet processing (see later).

Table 1. Comparison of fibre properties between PLA and PET [5].

Fibre properties	PLA	PET
Specific Gravity	1.25	1.39
T_m (°C)	130-175	254-260
Tenacity (g/d)	6.0	6.0
Elastic recovery (5% strain)	93	65
Moisture Regain (%)	0.4-0.6	0.2-0.4
Flammability	Burn 2 min after flame removed	Burn 6 min after flame removed
Smoke Generation	63 m^2/Kg	394 m^2/Kg
Limiting Oxygen Index (%)	26	20-22
Refractive Index	1.35-1.45	1.54

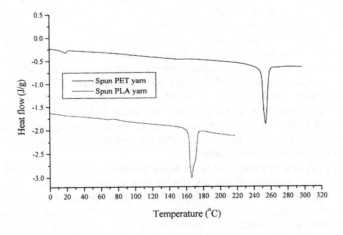

Figure 1. DSC scan of PET (top) and PLA (bottom) spun yarns.

Fibres of PLA have been successfully dyed with disperse dyes, although early practical experience suggested that the use of higher temperatures or longer times of dyeing could cause some degradation of the PLA. DyStar recommend dyeing conditions of 30 mins at 110°C and pH 4.5-5.0 (Figure 2.), the dyeing temperature being 20°C lower than that for PET fibres. At the same percentage depth of shade applied, a deeper shade is usually achieved on PLA fibres than on PET fibres of similar dimensions (dtex/filament), an exception being when (high wet fast) benzodifuranone dyes are used.

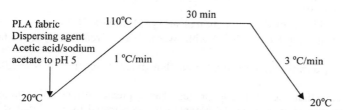

Figure 2. Disperse dyeing procedure used for PLA fibres.

The wet fastness properties of disperse dyes on PLA are slightly lower than those on PET, after a heat treatment (30 secs @130°C), designed either to convey dimensional stability to the fabric or to apply a chemical finish. A dyed Ingeo fibre fabric exhibited wash fastness properties, which were lower by 0.5-1.0 grey scale units than those of a PET fabric, dyed to the same visual depth of shade (and heat set for 30 secs @180°C). This is attributed to the higher degree of thermal migration of the disperse dyes to the surface of the Ingeo fibre, after heat treatment.

Over the last decade, large amounts of research have been directed into defining the optimum conditions for manufacturing and wet processing (dyeing) 100% PLA fibres.
More recently, work has been directed towards the production and evaluation of blends of Ingeo and natural fibres. Blending of PLA with cotton would result in a host of desirable properties such as improved moisture regain, better next-to-skin comfort etc. However, before such a blend can be successfully commercialised, the preparation and wet processing requirements of the two fibres need to be compatible. For example, the preparation (eg. scouring, bleaching) and reactive dyeing processes required for the cotton component in the blend are generally conducted under warm alkaline conditions, the type of conditions which might reduce the strength of the PLA fibre component in the blend.

Consequently a study was carried out to examine the reduction in fibre/fabric strength (estimated by measurement of molecular weight (M_n) and ball burst strength values, respectively) as a result of typical wet processing conditions.
  (a) Scouring but not bleaching (the cotton component) before high-temperature dyeing and 'reactive' dyeing.
  -Drop in M_n during alkaline scouring (66,000 → 64,000)
  -Drop in M_n during high-temperature dyeing (30 mins@110°C, pH 5)
    (64,000 → 57,800)

-Drop in M_n during dyeing (50°C) with 'warm dyeing' reactive dyes (57,800 → 53,600)

-Drop in M_n during dyeing (80°C) with 'hot dyeing' reactive dyes (57,800 → 54,400)

Since it is considered that a fabric blend would be sufficiently strong for commercial use if $M_n \geq 40,000$ (ball burst strength ≥ 50 lb), the above results indicate that defined processing route does minimal damage to the strength of the blend.

(b) Scouring followed by cotton bleaching (plus high-temperature dyeing and 'reactive' dyeing)
- Drop in M_n during alkaline scouring (66,000 → 64,200)
- Drop in M_n during peroxide bleaching (64,200 → 63,100)
- Drop in M_n during high-temperature dyeing (30 mins@110°C, pH 5) (63,100 → 51,900)
- Drop in M_n during alkaline reduction clearing (51,900 → 51,000)
- Drop in M_n during dyeing (50°C) with 'warm dyeing' reactive dyes (51,000 → 50,000)
- Drop in M_n during dyeing (80°C) with 'hot dyeing' reactive dyes (51,000 → 49,500)

The results indicate that the bleaching process has a greater effect on reducing the strength of the fabric than simple scouring, a conclusion confirmed by ball-burst strength data on the fabric and SEM data. Nevertheless, provided that excessively long bleaching times were not used, both M_n and ball-burst strength values remains above the critical threshold for commercial acceptability.

CONCLUSIONS

PLA fibre possesses promising properties as a textile fibre when compared with the conventional polyester (PET) fibre. It can readily be dyed with disperse dyes at 110°C, generally producing shades of similar depth to those on PET fibres. However the fastness properties (both light and wet fastness after heat treatment) of PLA fibres are generally inferior to those of PET fibres. The use of PLA fibres in the textile industry is currently being expanded via the development of blends with natural fibres such as cotton and wool. Ingeo/cotton blends provide a desirable combination of the properties of both fibres. It has been shown that the blend can be processed under commercially realistic preparation and dyeing conditions, without a deleterious effect on the strength of PLA fibre in the blend.

REFERENCES

1. S Jacobsen, Ph Degée and H G Fritz, 'Polylactide (PLA)- a new way of production', *Polym Eng Sci*, 1999 **39**(7) 1311-1319.

2. H R Kricheldorf, 'Synthesis and application of polylactides', *Chemosphere*, 2001, **43** 49-54.

3. P Gruber and M O' Brien, in Biopolymers, Vol. 4: polyester III -Applications and commercial products, (Weinheim: Wiley-VCH, 2002) 235.

4. J S Dugan, *INTC 2000*, Dallas, Texas, USA, 2000.

5. Information from Cargilldow- www.cargilldow.com.

ECO INITIATIVES IN THE TEXTILE PIPELINE – A SOUTH AFRICAN EXPERIENCE

[1]Pat Foure, [2]Tembeka Mlauli

[1]Clothing & Textile Environmental Linkage Centre, 15, Lower Hope Road, Rosebank, Cape Town 7700 South Africa
[2]Clothing, Textile, Footwear & Leather Sector, Trade & Investment South Africa, Department of Trade & Industry, P.O. Box 902, Groenkloof, Pretoria 0027 South Africa

INTRODUCTION

Although small by international standards, the South African clothing and textile sector has the potential to contribute significantly to South African export income and has been identified as one of eight key Industrial sectors with the greatest growth potential and marketability. Exports currently produce under 4 billion rand from a total production value in the region of 22 billion rand.[1] As it is a manufacturing sector under increasing pressure from external suppliers, particularly from China and India, strategies have been put into place by central government to work towards growing exports and slowing the growth of imports.

Increased environmental awareness and knowledge for the businesses in this sector together with improved environmental performance have been identified as important contributors to this strategy. Cleaner production initiatives have the duel advantages of improved economics in production and a better marketing profile and can therefore contribute to the growing of exports in the sector. In addition, the effects of invisible trade barriers relating to environmental management requirements can be minimized.

We will review the environmental initiatives and results of the Danish Cleaner Textile Production Project. Together with excellent examples of cost and resource savings, this project achieved additional positive outcomes including improved communications channels between the organizational role players and empowerment of employees working on environmental improvement projects.

In addition we will present an overview of some of the local technical research projects contributing to opportunities for environmental improvements in the sector. A range of locally grown fibres has been converted into innovative yarns and fabrics which meet industrial and geotechnical needs or have the potential to appeal to customers involved in fashion and design. We will also look at government initiatives to promote increased export opportunities and increased environmental responsibility by the manufacturers in this sector.

Finally, we will look briefly at the plans for future assistance to the South African clothing and textile sector in order to maintain the momentum in environmental improvements gained by the current activities.

CLEANER PRODUCTION IN COTTON GROWING

As the cause of major environmental impacts in the textile production pipeline, cotton growing was a focus area of the Danish Cleaner Textile Production Project running in

South Africa from 2000 to 2003. Using an entomologist and a cotton expert, the current polluting farming methods were reviewed. Existing cleaner cotton farming methods were available from the experience of small scale farmers who were unable to fund extensive inputs of pesticides, herbicides, fertilizers and defoliants as utilized by the commercial farmers.

Activities and achievements

The activities of the Cleaner Textile Production Project, Cotton Component covered feasibility studies, Integrated Pest Management Training (IPM) and the development of training aids with pegboard scouting techniques, which could be used easily even by illiterate small scale farmers. In addition farmer study groups were formed, a study tour for farmers to Denmark and Turkey organised, a highly experienced and respected cleaner production development adviser appointed, and the results of research and experience in the fields of IPM and organic cotton marketing were published.

Improvements were made through the IPM, scouting, demonstration projects, research and training and the dissemination of information. Activities included the study tour to Turkey and Denmark for 16 farmers and specialists, training for 185 people, research in 9 projects into cleaner production in cotton growing, and alternatives to the widely used GM cotton grown in South Africa. Four conferences and seven workshops were held to disseminate information within cotton farming. Publications included videos, posters, flyers, scout guides, magazines, newspapers, CD's and a website [2]

Check scouts were trained to assess insect populations and using research information on population sizes of pests which are known to cause crop damage, advice on the timing of chemical spraying was given. This Integrated Pest Management (IPM) was estimated, by participating farmers, to have resulted in yield increases of 15 – 20 % and savings of two crop sprayings, with estimated savings of R300/ha. Spray quantities were reduced because IPM enabled the crop to be sprayed when pests were at the larval stage, not adult. This attention to detail with the IPM techniques also resulted in improved farm management.[3]

The main constraints to further achievements were the resistance to change of both commercial and small-scale farmers and the insufficient financial incentives and research data to support further changes to organic or cleaner cotton growing techniques.

Cleaner Cotton Production Unit

In order to support ongoing initiatives in this area, the Cleaner Cotton Production Unit (CCPU) has been established, under the auspices of Cotton SA and managed closely with other cotton farming training initiatives. The objectives of the CCPU are to further support and strengthen awareness of Cleaner Cotton Production in the commercial and small-scale cotton growers in South Africa by
   – Promoting Integrated Pest Management (IPM) amongst Cotton growers in order to strengthen and sustain the concept of Cleaner Cotton Production
   – Strengthening awareness of cleaner production and specifically Best Management Practices & IPM, in order to expose the small-scale farmer to opportunities to enable him to consider organic or IPM as a viable option for future farming

The activities of the newly formed CCPU will include training programmes in all regions, dissemination & publication of information, demonstration projects in IPM on one commercial and four small-scale farms and check scout trials. To date their activities have included a newsletter distributed to the farmers, CDs distributed to farmers to inform about organic cotton production, a Cleaner Cotton Production Guide on laminated posters produced and distributed, articles on IPM in Farmers Monthly, three check scouts working on three demonstration projects each, commercial demonstration trials and an information day on cleaner production in cotton farming for both commercial and small scale farmers.

Continued support for these initiatives is required from Cotton SA in order to further promote the provision of information on cleaner production methods as an alternative to the intensive and polluting methods historically used by South African cotton farmers.

CLEANER PRODUCTION IN TEXTILE MANUFACTURE

The other major contributor to the environmental impacts from the textile production pipeline is the textile manufacturing sector with particular emphasis on the dyeing operation. This therefore was the other focus area of the Danish Cleaner Textile Production Project.

Activities and achievements

A wide range of technical activities were actioned including 4 study tours to Denmark, involving 40 key role players in the industry and related services. Also technical training for 106 people, 27 factory audits, 4 compressed air audits, 16 feasibility studies and 16 demonstration projects were completed. Dissemination activities included 4 conferences, 6 management awareness seminars, and 7 presentations, 16 workshops to industry groups on cleaner production activities, 5 training seminars and many publications including magazines, videos, CDs and a web-site.

A Cotton SA merit award has been competed for annually by textile manufacturers for the past three years. The award honours the company with the most environmental improvements and is hotly contested by the South African textile industry and used in marketing literature by successful companies. A past winner of this award won the inaugural national "Greening the Future Award" run by the Mail and Guardian newspapers and open to all South African Industry. This was considered an especial honour taking into account the historically polluting nature of the textile manufacturing industry.

Improvements in the environmental performance of Textile manufacturing in South Africa were made through questioning normal practices, thinking creatively, improving basic housekeeping issues, investing in technology, and catalytic in much of this change was the provision of waste minimisation training to the textile industry.

Cleaner Production results

Annual savings reported from Cleaner Production activities by participating companies as at June 2003 are detailed in Table 1.

Item	Annual financial saving	Annual unit savings
Water and effluent	4 970 000 (12)	790 000 kl (8)
Steam	2 560 000 (6)	31 000 tons (2)
Energy	565 000 (3)	
Heavy fuel oil	1 570 000 (3)	480 kl (2)
Chemicals	4 600 000 (6)	7 tons (1)
Waste	1 080 000 (4)	
Other*	3 660 000 (5)	
Total	19 005 000 (16)	

Table 1 : Reported annual savings from textile manufacturers participating in the Cleaner Textile Production Project [4]
* Other includes aspects such as raw material and consumables
The figures in brackets are the number of companies reported in this table

In addition other positive outcomes were experienced by participating organizations including increased awareness at industry level, increased awareness at local government level, improved company to company interaction, improved company to regulator interaction, and increased awareness at tertiary education institutions. This was achieved through the training, study tours, workshops, and publications,

It is encouraging to note that a portion of these savings have been used by several companies towards the costs of improvements in their environmental management and in the reduction of environmental risks.

There is, however, still enormous scope for additional cost and resource savings within the South African textile manufacturing industry, both in terms of further savings from companies already embarked on cleaner production and waste minimization programmes and from companies who had not been involved in these projects to date. The realization of these opportunities needs to be the key objective of current projects in this sector, such as the UNIDO-funded National Cleaner Production Centre.

Clothing & Textile Environmental Linkage Centre

To support the momentum created by the Cleaner Textile Production Project, an extension was granted with the funding by the Department of Trade and Industry together with the Danish Government of the Clothing & Textile Environmental Linkage Centre from 2003 to 2005 and the Clothing & Textile Environmental Linkage Centre was created. The objectives of the Centre include
-        Establishment and maintenance of a comprehensive repository of environmental information relating to the textile pipeline including the clothing and retail sectors.
         Raising awareness of environmental issues in decision makers throughout the textile and clothing pipeline.

Increasing the knowledge of environmental issues and how to integrate these into product requirements

Identifying and communicating market opportunities for textile products produced in an environmentally responsible manner.

Further details of the ongoing activities of the Clothing & Textile Environmental Linkage Centre are covered in later sections of this paper.

CLEANER PRODUCTION IN CLOTHING MANUFACTURE

In order to raise awareness of environmental improvements in the clothing manufacturing sector a Danish clothing consultant was utilized, to complete walk-through audits in four South African clothing manufacturers during March 2004. The major area of focus was the cutting department where the elimination of excessive fabric wastage can contribute economically and environmentally to a company's performance. Simple low cost improvements were recommended, suitable to the short run, labour intensive nature of the clothing manufacture in the medium sized manufacturers audited.

The observations and results of the audits were communicated to the companies involved with emphasis on potential cost savings from reduced wastage, and the inherent water, energy and other resource savings from reduced fabric usage. Typically, in many South African manufactured textiles, the resource savings when 100 kgs of fabric are saved can amount to between 6,000 and 10,000 litres water, with chemical and salt savings of 30kgs. In a water scarce country, with the inherent risks of salination of agricultural land, these savings can provide strong additional drivers for government support of cleaner production initiatives, from national, provincial and local water and agricultural authorities.

As expected, savings in fabric usage of over 2% were considered achievable with improved care and management of the basic processes within the cutting operation, together with some small investments. The implementation of many of these recommendations is currently in progress.

In order to build capacity, and facilitate skills transfer, South African skills training facilitators and cleaner production specialists participated in the audits, and will develop local training programmes in a similar manner to the highly successful waste minimisation training developed in the textile manufacturing sector.

The results of the completed audits serve as a demonstration to other South African clothing manufacturers of the opportunities for cost and resource savings and the specific recommendations and potential resultant savings are communicated throughout the industry via trade journals and workshops at tertiary education and technical training establishments.

ECOLABELLING & ORGANIC PRODUCTS

In order to demonstrate the ability of many South African textile and clothing manufacturers to meet ecolabelling criteria, a range of locally produced T-shirts were certified with the EU flower label for display at the World Summit in Johannesburg in September 2002. During 2003 several other manufacturers investigated certification with the EU flower and found that their operations and methods used would be capable of certification, should this be a requirement from a potential European customer.

Many manufacturers with current European export markets are already certified with the Oekotex label, with its criteria for their finished products. Certification with the wider life-cycle criteria of the EU flower is only allowed for specific products going into Europe and therefore has been limited in use to date.

The development of more production facilities with these certifications will remain demand driven at all times, both from the reasons of the rules for the label, and from the costs of certification where it is not a requirement by the customer.

Organic cotton production

Although South Africa has a large cotton growing industry, which at its peak volume could satisfy current local demand, there is no organic cotton grown in South Africa. Research was completed in 2003 into the production and marketing of organic cotton within the region, and it was concluded that as organic cotton lint is available from elsewhere in the sub-Saharan region, there is no immediate pressure to convert local production to organic.

Following a study tour to Turkey, one farmer in the Limpopo area has produced two crops of cotton from a small area of his farm, without the use of any chemicals on the crops. A pilot project has now been initiated to run trials in growing organic cotton in South Africa, and is currently at the planning stage.

The widespread promotion and use of genetically modified cotton by seed suppliers operating within South Africa remains an obstacle in the development of cotton production with improved environmental processes.

ENVIRONMENTAL MANAGEMENT SYSTEMS

One area encouraged by both South African National Government and European funders of development and environmental projects has been the certification of South African companies to international standards. There is widespread quality certification under ISO 9000/2000 within the textile industry, but environmental management system certification under ISO 14001 has been completed by only a handful of textile manufacturers. Those completing environmental certification are exporters who have found this to be a basic requirement when operating within the European markets.

Several textile companies are very close to achieving certification, but the barriers to completion are generally the costs of the auditing and certification procedure and the costs of administration of the systems. A further constraint is the shortage of skilled personnel resources at the required levels in these organizations to effect and maintain these management systems. Some of the physical improvements for risk reduction in environmental management can be costly and will generally be implemented over a prolonged period to comply with budget restrictions.

In general South African textile and clothing manufacturers are capable of achieving international standards and of obtaining certification, and this is achieved as soon as sufficient pressures are felt by the company from their potential International, and especially European customers.

NATURAL FIBRES

With its wide variations in climatic conditions there is enormous scope within the Southern African Region for the development of the production of a wide range of natural fibres for textile production.

The government supported, Centre for Scientific and Industrial Research (CSIR) has a specialist Centre for Fibres, Textiles and Clothing. In addition to technical support to the Industry, one of their key activities is the development of small scale and later commercial ventures in the growing, conversion and production of new textile products using natural fibres. Development projects in growing and producing textiles with hemp, sisal, flax and other plants have produced exciting and attractive fabrics, which are in varying stages of conversion to commercial ventures.

Key to this development has been a specially designed machine which "cottonises" the natural fibres by reducing the length and thickness of many of these fibres to a suitable length and tex for yarn and fabric manufacture. These shredded fibres can then be successfully blended with cotton to produce user friendly fabrics with a softer feel.

The large cactus like plant, S. Ethiopica has strong fibres in its leaves and these are being successfully converted to non-woven geotechnical fabrics. Work has started on the conversion of pineapple leaves into an attractive fashion fabric.

Wild silk project

Cocoons can be found on the camelthorn and mopane trees throughout a broad band of the region across Namibia, Botswana, Zimbabwe and the North West province of South Africa. A development project has therefore been initiated whereby the cocoons are harvested sustainably by the local population, when the moth larvae have left the cocoon. The cocoons are cleaned and degummed within the community and currently yarn is spun and fabric made in the CSIR research facilities in Port Elizabeth.

After completion of training, production will be moved this year to the North West province and production volumes will be increased to provide for commercial marketing and sale of these fabrics by the community. The wild silk fibres are blended with cotton to produce a beautiful, yet robust fabric with a potentially successful international niche market.

This project has many desirable aspects, with its community involvement, the utilization of a previously unused resource and long term sustainability as a result of the harvesting only of empty cocoons. The attractive, unusual fabric produced with its "Cinderella" fairy story should appeal to European and other export markets.

REMOVING TRADE BARRIERS

The South African government has been instrumental in negotiating preferential market access agreements with many key trading partners. The South African Free Trade Agreement with the European Union went into effect on 1st January 2000 and reached zero tariffs for many textile and apparel items by the end of 2003. The Africa Growth and Opportunity Act (AGOA) between South Africa and other sub-Saharan countries with the US took effect in March 2001 for apparel items and remains in force until September 2008.

The SADC trade protocols within the Southern African development community came into effect in September 2000 and provide for the phasing down of tariffs for 11 of the 14 SADC member countries. The participating member countries include a combined population of 135 million and aim to establish a free trade zone in this sub-Saharan region.

These preferential market access agreements can provide the platform for increased exports to developed markets, in the US and Europe and it is expected that these additional export opportunities will lead to further improved environmental management by South African companies from two major factors.

Firstly the pressures for responsible environmental management will be exerted by the increase in this environmentally concerned and aware customer base. Secondly the improved liquidity of the South African companies succeeding in this export market will enable them to budget expenditure on environmental improvements and construction for environmental risk reduction which had not been considered essential priorities when trading predominantly within the local market. So a combination of the need to improve will be combined with the accessibility of additional funds to improve environmental performance.

IMPROVING THE KNOWLEDGE AND AWARENESS OF RETAILERS, DESIGNERS AND BUYERS

One important means to provide ongoing pressure for improved environmental performance in the textile manufacturing industry is increased environmental demands from designers, buyers and retailers in the sector. The promotion of awareness and knowledge throughout the pipeline was therefore targeted as a key objective for the Clothing & Textile Environmental Linkage Centre. The ability of specifiers in the sector to be able to make informed choices can impact positively on the environmental performance of the clothing and textile manufacturers.

Students of Fashion and textile design have been targeted as necessary in providing for an informed future generation of designers in the industry. An Environmentally Friendly Design Competition was therefore launched this year, with workshops covering 485 students and 28 lecturers at 15 design departments in 8 tertiary institutions. 128 individual and group entries have been received to date and the 31 shortlisted entries will display their creations at the high profile Cape Town Fashion Week in August. Students from the winning groups will receive the prize of study tours to either the Danish Design School in Copenhagen or the Danish textile industry, Teko Centre in Herning. Already the levels of awareness concerning environmental considerations in lecturers and students of fashion and textile design have been substantially raised, and further promotion of the competition entries and winners by the media will continue this process further.

Fashion design is considered by the Department of Trade and Industry as one of South Africa's strengths and a key area for development, and it is important that our young designers understand the demands of their customers internationally regarding environmental considerations in their designs.

Retailers of clothing and yextile products can contribute enormously to the marketing and promotion of products with good environmental profiles. In South Africa, the leading retailer in this area is Woolworths, from the Marks and Spencer's stable. They have been leaders in the promotion of organic food in their food halls and have recently introduced a small trial range of organically produced cotton garments for children and ladies. Their influence over the local manufacturing industry in the clothing and textile sector is enormous and we have been lucky to have had the benefit of their input and committed support as members of the Steering committee for the Cleaner Textile Production Project and for subsequent initiatives. Other South African retailers, like the large Edcon group are starting to look at their environmental profile and investigating staff training and the reporting of environmental performance.

Another area to be targeted under the continuing Cleaner Textile Production inititative is the government buyers through tender requirements and tender adjudication procedures.

Currently the performance of potential suppliers is assessed on social and black empowerment criteria. There is an objective to include environmental criteria in these purchasing processes.

ENVIRONMENTAL LEGISLATION AND ENFORCEMENT

During the last ten years, there has been a revision of environmental legislation and many new, excellent, forward thinking laws have been passed. The environmental rights of all citizens have been entrenched by inclusion in our new constitution and the National Environmental Management Act together with separate water management and other legislation provides a very strong basis for good environmental governance at national, provincial and local levels. The inclusion, in legislation, of requirements for environmental impact assessments to be performed in advance of proposed developments, should ensure that any future developments are planned on a sustainable basis.

However enforcement levels have lagged behind this legal framework and hampered effectiveness of the legislation in several areas. This has been due, in part, to the initial lack of capacity to provide the enforcement and in part to the conflict in some areas between the provision of employment and the means of enforcement. Despite the willingness of government agencies to improve enforcement the priorites of government must remain on the provision of jobs and promotion of economic development to enable acceptable basic living conditions to be enjoyed by all South Africans, and improve the standard of living of the previously disadvantaged majority in the country.

Innovative solutions to these dilemmas are therefore required, and the advancement of cooperation between regulators and industry is important in these solutions. The Cleaner Textile Production project included both industry and regulators in study tours and cleaner production training with the objective of enabling win-win solutions to be found when regulators can work together with industry in assisting them to lessen their environmental impacts. Methods such as the "score system" where regulators and industry can together analyse the content and effects of industrial effluent on a long term basis instead of spot check sampling with its inherent conflict potential can contribute to achieving the required co-regulation.

A guide in cleaner production in the textile sector has been produced especially for regulators and this has been distributed free of charge to the relevant departments together with training in the use of the guide. Ongoing efforts, in particular from the regulators, are essential to continue this process and to reach a stage where our good legislation can become effective. This will need driving with external input to achieve these desired goals within a reasonable time scale.

SOUTH AFRICA AND THE NEIGHBOURING STATES

South Africa has been one of the founding partners of the new partnership for Africa's development, NEPAD, which provides a united platform for the advancement of African development. The goals of this initiative have gained support from many key northern partners including the UK's Development Ministry, so that Africa has been placed at the forefront of the UK development agenda.

NEPAD provides a vision and strategic framework for Africa's renewal and among its priorities is good economic governance, regional co-operation and capacity building. Two of its priority sectors for policy reforms and increased investment are the environment and market access. Several identified objectives in the manufacturing sector relate to the promotion of international standards and certification within the region as a support to the growth of trade with the developed world.

Improved communication and closer ties between the sub-Saharan countries in the last ten years have gone a long way to removing the barriers of distrust of South Africa caused both by the previous political base and the ongoing situation of the size and economic muscle of South Africa relative to many of its neighbouring states.

Many of the negotiated preferential trade agreements have included neighbouring states and this achieves greater unification of the region and stronger shared goals for development and responsible governance. South Africa has been able to offer considerable technological assistance to neighbouring states and this assists with strengthening the economic base in the region in order to provide a stronger platform for improvements to environmental responsibility.

There are several regional initiatives in the textile and clothing sector, including the Sustainable Trade and Innovation Centre, Southern African Regional Consultation Workshop which was hosted in South Africa in December 2003 with support of Department of Science and Technology. This organization promotes trade and market access in sustainable goods and services between developed and developing countries and is supported in the sub-Saharan region by South Africa.

ONGOING INFORMATION AND SUPPORT FOR ENVIRONMENTAL IMPROVEMENTS IN THE SOUTH AFRICAN CLOTHING & TEXTILE PIPELINE

The current life spans of both the clothing and textile Environmental Linkage Centre and the National Cleaner Production Centre continue until the end of 2005. The Department of Trade and Industry, both in the Trade and Investment, Clothing & Textile Section and in the Environmental Section are committed to continuing to provide support for this key Industrial sector in aspects of environmental improvements. Firm agreement on what form this support should take, and under which umbrella it should be located, has yet to be confirmed, but is expected to be finalized early in 2005.

Further support for the ongoing provision of environmental support to the clothing and textile sector will come from all other interested and affected parties from within the three spheres of government, industry, retailers, specialists and organized labour. The steering committee which has guided the Cleaner Textile Production Project, the Clothing & Textile Environmental Linkage Centre and the National Cleaner Production Centre contains representatives from these spheres and this has improved the knowledge, understanding and commitment of these key players to environmental improvements in this sector.

CONCLUSIONS

The ability of many South African Textile manufacturers to meet European and US standards is evident. The main barrier to continued environmental improvements is the lack of pressures on manufacturers to embark on further expenditure on these improvements.

The responsibility for providing this ongoing pressure will rest mainly with government agencies, who together with increased consumer knowledge and demand can continue to fuel the process. Opportunities for exports, particularly to the European markets would create further demand pressures on industry, for environmental improvements including more formal certification. In addition, where savings can be achieved from waste minimization and other cleaner production efforts, this will offset, to an extent, and provide additional motivation, for further in-house improvements with net costs rather than net savings. The need for ongoing assistance and support from a central, efficient and economically effective resource base will continue if these environmental improvement objectives are to continue to be realized.

South Africa is currently a country experiencing a wealth of positive changes throughout government, business and civil society, and it is the belief of the authors that this increased   flexibility and open-mindedness, together with national government commitment for improvement, will continue the process of reducing the environmental footprint of our clothing and textile industry.

REFERENCES

1 E Krzysztof, S A Wojciechowicz, 'Manufacturing Trends 1993', Statistics South Africa, *SA Customs & Excise, Department of Trade and Industry*, South Africa, 2004

2 K Lundbo, Draft Completion Report, Cleaner Textile Production Project South Africa, Darudec, *Royal Danish Ministry of Foreign Affairs*, June 2003

3 J Hanks and C Janisch, Evaluation of Cleaner Production Activities in South Africa, Danida, *Royal Danish Ministry of Foreign Affairs*, June 2003

4 S Barclay, 'Cleaner Production makes good business sense: Results of the Cleaner Textile Production Project', *Water Institute of South Africa*, Biennial Conference, 2004

ACKNOWLEDGEMENTS

Danida, Royal Danish Ministry of Foreign Affairs c/o Royal Danish Embassy, South Africa

Ms Karen Lundbo	Chief Technical Adviser	Cleaner Textile Production Project
Ms Susan Barclay	Textile Co-ordinator	Cleaner Textile Production Project
Ms Annette Bennett	Cotton Co-ordinator	Cleaner Textile Production Project
Dr Sarel Broodryk	Cotton SA	Cleaner Textile Production Project

Part III

Sustainable development and renewables

THE USE OF RENEWABLE RESOURCE BASED MATERIALS FOR TECHNICAL TEXTILES APPLICATIONS

Roshan Shishoo

Shishoo Consulting AB, Svartlavsvägen 18, Askim, Sweden

roshan.shishoo@shishooconsulting.se

ABSTRACT

The renewable resource policies on a global level play an important role for the sustainability of ecosystems. These are also vital for the sustainable growth of the textile industry. Materials from renewable resources, e.g., agro-cellulose fibres are becoming increasingly interesting for many technical textiles applications. Potential markets for agro-fibre based products include absorbents, geotextiles, filters, bio-composites for automotive and building sectors , packaging etc.

In this paper an overview will be given of the state of the art regarding renewable resource-based materials, the relevant processing technologies for the production of bio-composites, and the potential market areas.

Current drawbacks of natural fibres from the point of view of processing technologies and product performance will be discussed. In this context the use of plasma technology as an environment-friendly process in general, and for modifying the surface energies of agro-fibres in particular, will also be discussed. Other developments such as cellulose and starch-based biopolymers for use as matrices or melt fibres in technical textiles and bio-composites will also be discussed.

INTRODUCTION

As shown in Table I, the global production of natural fibres in the year 2002 amounted to over 35 million tons and the natural polymer fibres (man-made cellulosic fibres amounted to nearly 2.7 million tons. The quantities of flax, jute and ramie together constituted one third of the total production of the natural fibres.

In a number of consumer and technical applications of fibres, man-made cellulosic and starch-based fibres are seen to have a good long-term future as they are produced from renewable resources such as wood pulp and corn. For obvious reasons these type of fibres are also fully degradable. Until the introduction of lyocell and poly (lactic acid) or PLA fibres, viscose and cellulose acetate fibres made up the bulk of the natural polymer fibres produced.

Table I. Production statistics of man-made and natural fibres'
million tonnes *(2001 and 2002)*

Fibre	2001	2002	% change 01/02
Man-made fibres			
Cellulosic (viscose, lyocell etc)	2.67	2.72	1.6 %
Synthetic (polyester, nylon etc)	31.83	33.80	6.2 %

% of World Total Fibre Production in 2002 : 58 %

Natural fibres			
Cotton	20.07	20.62	2.8 %
Wool	1.40	1.35	- 3.1 %
Jute	3.35	3.22	- 3.8 %
Flax	7.12	7.20	1.3 %
Ramie	2.00	2.01	0.1 %
Silk	0.88	0.92	4.0 %

% of World Total Fibre production in 2002 : 42 %

Source : Fiber Organon July 2003

Natural fibres

Natural materials from renewable resources, e.g., agro-cellulose fibres are becoming increasingly interesting for applications other than apparel and household textiles. Because of their ecological advantages over oil-based polymers there are many possible applications for these materials in technical and industrial applications.

The great potential of using agro-fibres in the production of nonwovens, technical textiles and composites offer a new environment-friendly strategy for industrial and technical products. To properly utilise agro-fibres, it is necessary to have a good basic understanding of the property requirements of these types of fibres for various end uses.

Whereas agro-fibres as renewable resources are perceived to have ecological and economical advantages however, the factors such as fibre quality and supply are seen as bottlenecks in the industrial utilization of this fibres in technical textiles applications. Also fibre processing is regarded as complicated both technically and economically as is the relationship between price and performance.

Cellulose is the most abundant of naturally occurring organic compounds, being the chief constituent of the cell walls of higher plants supporting the structure of the plant. It constitutes at least one-third of all vegetable matter in the world. The cellulose content of vegetable matter varies from plant to plant. The botanical classification of the natural fibres is carried out according to part of the plant where they are found:

* bast- or stem fibres (e.g., flax, jute, ramie)
* seed fibres (e.g., cotton, kapok)
* leaf fibres (e.g., sisal, palms)
* fruit fibres (e.g., coconut)

The cell walls of these fibres consists of cellulose and lignin. Values observed for the tensile strength, modulus and elongation of natural fibres depend on their cellulosic content, microfibrillar angle, cell dimension, cell shape and cell arrangement.

The fibre dimensions and some structural, physical and mechanical properties of major types of natural fibres are given in Tables II-IV

Table II.   Structural and Mechanical Properties of Some Natural Fibres [1]

Fibre	Cellulose fraction	Angle	Cell length L (mm)	L/D-ratio	Tensile str. (Mpa)	Elong. (%)
Ramie	0.83	7.5	154.0	3500	870	1.2
Hemp	0.78	6.2	23.00	906	690	1.6
Jute	0.61	8.0	2.30	110	550	1.5
Flax	0.71	10.0	20.00	1687	780	2.4
Coir	0.43	45.0	0.75	35	140	15.0

Table III.   Dimensions of important natural fibres

Fibre	Fibres Length (mm)		Diameter (mm)	
	Min-Max	Average	Min-Max	Average
Flax	8-69	32	0.008-0.031	0.019
Jute	0.75-6	2.5	0.005-0.025	0.018
Ramie	60-250	120	0.017-0.064	0.040
Hemp	5-55	25	0.013-0.041	0.024
Sisal	0.8-7.5	3	0.007-0.047	0.021
Coir	0.3-1	0.7	0.010-0.024	0.020
Cotton	10-50	25	0.014-0.021	0.019

Table IV. Some physical and mechanical properties of natural fibres

Fibre	Density (g/cm³)	Tensile strength N/tex)	Breaking extension (%)	Work of rupture (mN/tex)	Initial modulus (N/tex)	Moisture absorption (%)	Effect of heat (100 °C)
Flax	1.54	0.54	3.0	18.0	18.0	7	stable
Jute	1.44	0.31	1.8	17.2	17.2	12	stable
Ramie	1.56	0.59	3.7	14.6	14.6	6	stable
Hemp	1.48	0.47	2.2	21.7	21.7	8	stable
Sisal	1.45	0.30	3.0		15.0	11	stable
Coir	1.24	0.31	1.8		17.2	10	stable
Cotton	1.52	0.19-0.45	5.6-7.1	3.9-7.3	5.9	7-8	stable

The density of natural fibres varies from 0.9 to 1.5 g/cm³ as compared to glass fibres of 2.5 g/cm³. Flax, sisal and hemp are most interesting and suitable for technical applications. Flax fibres have already being used in reinforcing applications of polymeric matrices and also for application in the automotive industry.

Flax is a stem plant and the fibre content of the dry stem (straw) is about 25%. Flax fibres have been used in textile fabrics for ages and the use of substitute for glass fibres, asbestos, reinforcement of polymers and papers is very promising. The flax fibre is a strand of cells, its thickness depends upon the number of these cells in any one fibre cross-section. it seems that about 3 to 6 cells or elementary fibres constitute a macro-flax fibre with thickness varying from 10 to 20 μm. Flax is more crystalline than that of cotton having a relatively high modulus and tensile strength make it very suitable for a number of technical applications. Flax also possesses a high heat resistance compared with many natural fibres.

The use of a renewable source based fibres in Europe can only be economically advantageous if fibre prices are low. Therefore only these fibres that are traded in huge quantities in Europe and hence are therefore available at prices considerably lower than the price for glass fibres are of interest. This important consideration means that sisal-, jute-, coconut fibres and flax tow are interesting, because they are available for about one fourth of the price of a glass fibre. However coconut - and jute fibres are not suitable for reinforcing purposes because of their low fibre strength. Hence flax and sisal are more suitable. A comparison of some of the mechanical properties of sisal-, flax-, and E-glass fibre is given in Tables V and VI.

Table V. Mechanical properties of sisal and flax fibre compared to glass fibres [2]

Property	Dimension	Sisal	Flax	E-glass
Tensile strength	MPa	610	900	2300
Modulus of elasticity	Gpa	28	50	73
Elongation	%	2.2	1.8	3.2
Density	g/cm³	1.3	1.5	2.6
Specific strength	Mpa/r	470	600	880
Specific stiffness	Mpa/r	22	33	28

Table VI. Composition of flax - and cotton fibres [3]

Fibre	Components with respect to the dry material (%)						
	Cellulose	Lignin	Waxy subst.	Pectines	Nitrogen	Ash	Pectose
Flax	72.9	4.6	2.1	3.7	0.35	0.86	3.30
Cotton	84.3	1.87	1.12	1.21	0.26	1.00	1.73

NATURAL POLYMER FIBRES

Solvent-spun cellulosic fibres – Lyocell

There are several solvent systems known to be effective for the direct dissolution of cellulose, for example: lithium chloride/ dimethyl acetamide; ammonia/ amonium thiocyanate. The former Courtaulds company in the UK developed a fibre spinning process based on the use of amine oxide, N-methyl morpholine oxide, to effect the dissolution. This compound is non-toxic and therefore very attractive for the basis of a manufacturing process. Furthermore, the properties of the fibre produced show advantage over cellulosic fibres produced by other processes.

The lyocell spinning process is illustrated in Figure 1. Wood pulp and amine oxide (N-methyl morpholine oxide - NMMO), as a solution in water, are mixed and then passed to a continuous dissolving unit to yield a clear, viscous solution. This can then be extruded into a dilute aqueous solution of the amine oxide, which precipitates the cellulose to fibre. After washing and drying, the fibre is ready for processing. The diluted amine oxide must be purified and is then re-used after removal of excess water. Thus the process utilises materials which are environmentally clean, and recycling of the solvent is an integral part of the process. Waste products are therefore both minimal and non-hazardous, and control is very easily managed.

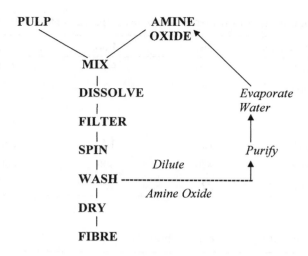

Figure 1 Lyocell fibre process outline

Lyocell fibres initially produced by Courtaulds, UK under the trade name Tencel sold their business to Lenzing, Austria who also manufacture under the trade name Lenzing lyocell. NewCell lyocell fibres are produced by Akzo Nobel.

In Table VII some structural characteristics of viscose fibre are compared with those of solvent-spun fibres. In Table VIII some physical and mechanical properties of viscose fibre are compared with the corresponding properties of Tencel lyocell fibre.

Table VII.   Structural comparison of viscose and NMMO fibres

Structural parameters	Viscose (normal)	NMMO (Lyocell)
Cross-section shape	Lobate	Round / Oval
Cross-section morphology	Core/Skin	Homogeneous/ Dense
Crystallinity	Variable	High
Crystallite length	Smaller	Larger
Crystalline width	Larger	Smaller
Crytalline orientation	High (lamella effect)	High
Amorphous orientation	Variable	High

Table VIII.   Some properties of lyocell (Tencel) and viscose staple fibres

Fibre property	Viscose	Tencel
Titre (dtex)	1.7	1.7
Strength, cond.(cN/tex)	26.0	42.0
Strength, wet (cN/tex)	13.0	38.0
Elongation, cond. (%)	18.5	11.0
Elongation, wet (%)	21.0	13.0

The tenacity of Tencel especially the wet tenacity is higher than that of the other man-made cellulosics. Also of significance is the very high modulus of the lyocell fibre. X-ray studies of the fibre have shown the cellulose crystals to be highly parallel in the longitudinal direction of the fibre. This no doubt contributes significantly to the high tensile strength of the fibre, and it also leads to the propensity for some degree of fibrillation, for example , during abrasion in the wet state.

One can say that lyocell, the new generic fibre type, is a cellulose fibre made by the best available technology to minimise environmental impact by virtue of both being based on the renewable resource and being processed using a non-toxic solvent and solvent extraction built-in process. Lyocell fibre has a higher average molecular weight, orientation and crytallinity than viscose rayon, resulting in high tenacity/ modulus. Lyocell retains most of its high strength and resiliency when wet.

Poly(lactic acid) fibre

PLA is an aliphatic polyester where the monomer is lactic acid. The lactic acid can be produced by a fermentation process using lactic acid-forming bacteria. The natural PLA polymer route to fibres and nonwovens is shown in the flow diagram below.

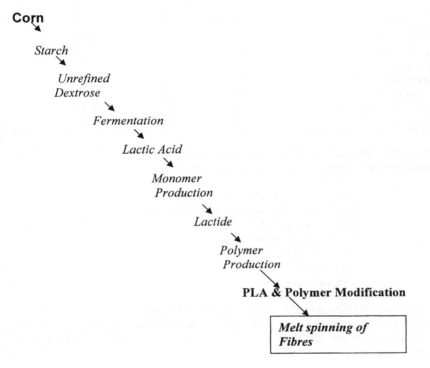

Corn

Starch

Unrefined
Dextrose

Fermentation

Lactic Acid

Monomer
Production

Lactide

Polymer
Production

PLA & Polymer Modification

Melt spinning of
Fibres

Two types of lactic acid,as two stereo-isomers (L and D-forms), i.e. L-lactic acid and D-lactic acid, can be produced by natural fermentation and from these monomers are obtained L-lactide, meso-lactide and D-lactide. Optical isomers added can allow L/D ratios to be controlled from 99%L to 80%L grades.

The morphology of the polymers and copolymers of lactic acid can be varied from amorphous to crystalline. Copolymerization of e.g. L-lactide with caprolactone, leads to materials with a wide range of properties from glassy to rubber-like. The tensile strength of these copolymers can vary from 0.6 to 48 MPa whereas the elongation varies from 1 to approx. 400% oriented film and fibres have a tensile strength at least 10 times higher.

The physical properties of poly (lactic acid) filament is between those of PET and PA6 (Table IX)

Table IX.   Physical properties of PLLA, PET and PA6 filament yarns [4]

Properties	PLLA	PET	PA6
Tensile strength			
(g/dtex)	4.0-5.5	4.0-5.5	4.0-5.5
(Gpa)	0.40-0.55	0.44-0.61	0.36-0.5
Elongation (%)	20-35	20-35	20-35
Young's modulus			
(g/dtex)	60-70	90-120	20-40
(Gpa)	6.0-7.0	10-13	2.7-3.6
Density (g/cm^3)	1.27	1.40	1.15
Crystallinity (%)	83.5	78.6	42.0
Refractional index	1.45	1.58	1.53
Melting temp. (°C)	175	256	222
Crystallizing temp. (°C)	103	170	140
Glass tr. temp. (°C)	58	69	50
Shrinkage in boiling water (%)	8-15	8-15	8-15
Moisture regain (%)	0.6	0.4	4-5

PLA fibres are now being commercially produced by many fibre companies. Some important consumer relevant attributes of PLA fibre are:

- Low moisture absorption & high wicking
- Low flammability and low smoke generation
- High resistance to UV
- Low index of refraction
- Lower specific gravity
- High elastic recovery
- High stain resistance

The above key performance characteristics of PLA fibres makes it very suitable raw material for many applications such as active wear, technical textiles, nonwovens, apparel fashion and home furnishings. PLA fibre can also be used as binder fibre in nonwoven production.

TECHNICAL APPLICATIONS OF NATURAL FIBRES

The price and environmental considerations are two factors which would favour the use of certain natural fibres. However there are some important problems such as uneven quality, continuity and consistency of supply and high moisture sensitivity which need to be solved before natural fibres will be able to replace many synthetic and glass fibres in technical applications. Also there is the need to change the existing negative image of agriculture as a supplier of industrial raw material.

The biggest single application today is the use of natural-fibres in automotive parts. But there are growing markets in other field such as medical, fibre-filled plastics, sport articles, boats etc. The existing and potential technical applications can be classified by use such as geotextiles, filters, sorbents, structural composites, non-structural composites, melded products and packaging.

The successful, commercially-viable production of elementary flax fibres would fulfil the material requirements by the producers of nonwovens, tissue, absorbent material, packaging material and fibre-reinforced plastics. Such a fibre would enable the end users to replace the currently used synthetic fibres and glass fibres. The advantages with natural fibres are that these are ecological, with low density and both low level of toxic emission. The disadvantages are mainly referring to their quality, namely irregular form, moisture sensibility, structural variations and thermal insensitivity to moulding processes. The objectives of major interest are :

- The possibility of developing flax elementary fibres capable of meeting the demands of the industrial partner interested in technical applications.

- Identification and evaluation of processing techniques to produce elementary fibres. The process has to be environmentally and economically sound.

- An overall conclusion regarding commercially viable flax fibre production processes and need of further developments.

Characterization of flax fibres in terms of dimensions and form, and a classification of elementary fibre content from fibre bundles after different defibrillation techniques. The studies were made using a video microscope as well as scanning electron microscope. Fibre damage was studied by scanning electron microscopy. The Video microscopic analyses and visual analyses of samples, were carried out for the determinations the overall result of the fibrillating process. Chemical modifications were made on the elementary fibres to meet the technical specifications, both in processing and of the end product. The main modification concerned wettability which was accomplished by using enzymes and/or wet bleaching processes. Different fibrillating processes resulted in elementary fibres with low wettability.

Modifications of the dispersability of fibres in a wet laid nonwoven process were also necessary. For liquid absorption end-users, the fibres tested in general had low wettability. Modifications were made by an ordinary wet bleaching process. It was also demonstrated that a plasma treatment could also increase the wettability of the fibre specimens.

Four processes of producing elementary flax fibres are:

- Steam explosion technique
- Sulphate process
- Extrusion technique
- Ultra-sound process

It is possible to obtain elementary flax fibres or bundles of fibres having the functional properties listed in specifications. Within the framework of this project the properties were not, however, optimized for the production of various types of end products. There is a need for further research and development in this area. However, two processing techniques were identified to give suitable elementary fibres. One was the sulphate process commonly used to produce cellulose fibres from wood and the other was a steam explosion process as used by IFTH, Lyon. These processes have the potential to be commercialized and there is a strong need for further development in order to optimize industrial treatments for producing elementary flax fibres for technical applications.

Some properties of elementary flax fibre are compared with other types in Table X.

Table X.   Comparative properties of some technical fibres [5]

Fibre type	Diameter μm	Density g/cm³	Breaking strength Mpa	Elastic modulus GPa	Breaking strain %
Polyester (high tenacity)	21 - 31	1.38	1100 - 1140	12 - 15	11 - 14
Glass (E-type)	5 - 24	2.6	2400 - 3400	73	3.8
Aramid	13	1.44	2700 - 3150	60 - 90	3.4
Flax (elementary fibre)	11 - 17	1.2	2000	85	2.4

Lyocell in nonwovens applications

The relatively high strength and high modulus of Tencel together with non-creep characteristics give significant advantages to materials for technical use, for example, in substrates for coatings. Tencel fibre can be used in a variety of nonwoven systems which use both dry and wet laid webs and utilize a range of bonding techniques such as latex bonding thermal bonding and hydroentanglement.[6]

Nonwovens produced from Tencel show very high tensile strength, particularly in the wet state, and are significantly stronger than those produced from viscose. This is the case also for both latex-bonded and thermal-bonded webs (Tables XI and XII).

Table XI.   Relative strengths of Tencel and viscose latex-bonded nonwovens

	40 g/m2 @ 20 % LATEX			
	Tencel, dry	Tencel, wet	Viscose, dry	Viscose, wet
MD strength	203	153	100	51
CD strength	172	119	100	57

Table XII. Tensile properties of 50/50Tencel/PP, and 50/50 viscose/PP thermally-bonded nonwovens.

| Bonding | | Tensile strength (daN) | | | |
| | | Machine Direction | | Cross Direction | |
		Dry	Wet	Dry	Wet
Plain	Tencel	9.2	10.9	2.6	2.5
	Viscose	7.0	6.5	1.1	1.0
Point	Tencel	7.2	8.2	1.6	1.9
	Viscose	5.2	4.1	0.7	0.7

FIBRE COMPOSITES FROM RENEWABLE RESOURCES

Bio-composites can ideally be defined as any combination of renewable resource-based fibre structures held together by a renewable resource-based matrix. The objective is to combine two or more materials in such a way that a synergism between the components results in a new material with much better properties than the individual components. At present many commercial agro-fibre based composites are made by using non-renewable matrix systems. Bio-composites can be classified in many ways: by their densities, by their uses, by their manufacturing methods, or by other systems. As such these are being classified as low-density / high-density composites, or structural / non-structural composites, or injection moulded / compression moulded products etc.

A structural bio-composite is defined as one that is required to carry a load in use such as in building industry, automotive industry and aerospace industry. Non-structural composites are not intended to carry a load in use and are used for products such as doors, windows, furniture, automotive interior parts etc. In some cases, one type of composite can be used for more than one use. For example, once a fibre web has been made it can be directly used as geotextile, filter, or sorbent, or it can be further processed into a structural or non-structural composite, moulded product, used in packaging or combined with other resources.

The most generally applicable agro-based fibres are flax, cotton, jute, sisal, straw and wood. The main criteria for the selection of plant fibres are price of the fibre, technical and chemical properties, performance and environmental aspects. Special quality criteria for agro-based fibres for use in bio-composites are: adhesive properties reinforcing potential (breaking strength, impact strength), stiffness, wear resistance, brittleness, moisture-related properties (ageing, form stability, swelling), heat stability, purity, resistance to micro-organism, non-odour, resistance to chemicals etc.

Resins and matrices

A wide variety of different synthetic matrices, both thermoplastics and thermosetting resins, which now generally are used for glass fibre reinforced composites have potentials for development of new composite materials with agro-fibre reinforcement and/or fillers. The most frequently applied thermosetting resins are urea formaldehyde, phenol-formaldehyde, melamine-formaldehyde, epoxy and unsaturated polyesters. Thermoplastics used generally are polypropylene, polyethylene, PVC, nylons etc.

In this context, the most interesting developments are those related to thermoplastic biopolymers. Of particular interest are cellulose acetate, cellulose propionate (CAP) and cellulose butyrate(CAB). The successful incorporation of biopolymer-based matrices would be a great step towards the realisation of true bio-composites.

An overview of possible applications of natural fibres in polymer processing is shown in Figure 2 and the technologies to produce fibre-reinforced composites made from natural-fibres and matrices from renewable resources are shown in Figure 3.

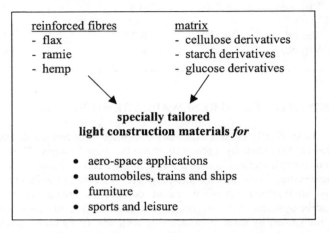

Figure 2 Fibre reinforced composites [7]

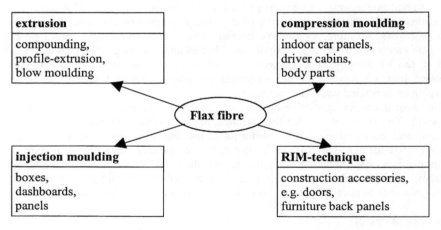

Figure 3 Overview of possible applications of flax in polymer processing [7]

Technology for high-density, thermoplastic composites: High density composites are those in which the thermoplastic component exists in a continuous matrix and the natural cellulose component serves as a reinforcing filler. The processes include *Compounding, Profile Extrusion and Injection Moulding.*

Problems using ligno-cellulose fibres in this area focus on the differences in bulk density of the fibre versus the polymer matrix components and the degree of shear of the compounding equipment, and thus fibre length retention or loss. Additionally fibre-matrix adhesion has to be enhanced

Technology for low-density thermoplastic products: Low matrix content composites can be made in a variety of ways. In their simplest form, the thermoplastic matrix acts much the same way as a thermosetting resin, that is, as a binder to the lignocellulosic component. An alternative way is to use the thermoplastic in the form of a textile fibre. Many synthetic polymer- based binder fibres are now commercially available and some chemical companies are also developing binder fibres based on biopolymers. The meltable textile fibre enables a variety of natural cellulose fibres to be incorporated into low-density,. nonwoven mat.The mat may be a product in itself, or it may be consolidated into a higher density product.

Nonwoven textile-type Composites

In contrast to high-matrix content composites and conventional low-matrix content composites, nonwoven textile type composites typically require long fibrous materials for their manufacture. Nonwoven processes allow and tolerate a wider range of cellulose materials and synthetic fibres depending on the application. After the fibres are dry blended, they are air-laid into a continuous, loosely consolidated mat. The mat then passes through a secondary operation in which the fibres are mechanically entangled or otherwise bonded together. This low density mat may be a product in itself, or the mat may be shaped and compressed in a thermoforming step.

Alternatively the long bast fibres can be used in place of glass fibres in many type of liquid composite moulding systems such as resin transfer moulding (RTM), resin injection moulding (RIM), structural reaction injection moulding (SRIM) and sheet moulding compounds (SMC). All of these techniques include a fibrous mat mixed with a liquid resin, which is polymerised to form a reinforced fibre composite. Agro-fibres are lower in specific gravity, higher in specific strength, lower in costs and less energy intensive to process compared with glass fibres, so they are well suited for these type of technologies.

Moreover, the main drawback with agro-fibres is the lower processing temperature needed because of the potential degradation of lingo-cellulose at elevated temperatures. For this reason the type of thermoplastic matrices or resin that can be incorporated with agro-fibres has earlier been limited to commodity polymers such as polyethylene (PE), polypropylene (PP) and PVC.

The effect of heat on the tensile properties and ageing of some natural fibres are shown in Tables XIII and XIV[8].

Table XIII. Tensile strength (cN/tex) of fibres before and after heat treatment

Fibre	Untreated	Fibres heat at (°C)			
		150 5 min	150 15 min	220 5 min	220 15 min
Fibres from Flax straw	38.6	44.0	47.5	34.6	34.5
Flax fibre after steam explosion	68.1	63.5	58.6	66.1	53.4
Ramie	61.4	65.5	58.8	42.1	27.8
Cotton	21.9	31.1	28.5	23.2	20.9

Table XIV. Tensile strength of flax and ramie fibres after steam explosion- ageing (cN/tex)

Fibre	Untreated sample	Aged sample	Change in tensile strength (%)
Flax	68.1	56.6	-17
Ramie	61.4	43.0	-30
Cotton*	39.1	41.6	+6

Development of renewable resources based material for automotive interior applications– Green Materials project

The results last year from a national project in Sweden called Green Materials, demonstrated that it is possible to obtain automotive interior panel materials solely made of renewable resources [9]. The work was focused on three different automotive applications; door panels, headliners and sound absorbents and have resulted in a number of small laboratory made materials but also a first lot of prototype materials.

The work has been concentrated on a couple of interesting materials. As renewable matrices there have been investigations based on polylactic acid (PLA), cellulose esters (CAB, CAP) and the aliphatic co-polyester, Eastar-Bio supplied by Eastman Chemicals. In the project fibres of different types of flex, sisal and wood fibres were used.

Investigations have been carried out in various laboratories across Sweden on the matrices, including rheology measurements, surface energy analysis, chemical composition analysis and compatibility studies of the interface matrix/fibre. The fibres too have been investigated in the same way with surface analysis using ESCA, and morphology, surface energy and surface charge measurements. Tests have been made on the derived composite materials namely impact strength, flammability and fogging.

The processes investigated include extrusion to make preforms or compounded materials. A tool to make compression moulded panels has been specified and a number of prototypes have been manufactured. A number of carded preforms, containing a blend of needled punched mats as preforms were manufactured and tested.

The most promising composites are natural fibres combined with PLA and plasticised CAB Most of the evaluations made so far shows promising results. The main drawbacks include odour problems and that the materials tend to be dense and in some cases too brittle.

Surface energies and adhesion

When a bio-composite fails by an interfacial or adhesive type mode, it is presumed that part of the failure arises from the lack of sufficient chemical bonding between fibre and matrix. Strength of composites will therefore improve if one can modify the nature of the component surfaces so that their surface energies are more compatible with one another.

The cellulose molecule is inherently hydrophilic and good wetting of cellulose fibres by any non-polar molten binder fibre or matrix in a bio-composite is essential for the composite manufacturing process. Several different types of functionalized additives have been used to improve the dispersion and the interaction between cellulose – based fibres and polyolefins. In the case of polypropylene/cellulose composites, maleic anhydride (MA)-grafted polypropylene (MAPP) has been reported to significantly improve the bonding between agro-fibre surface and the PP-backbone chains. The maleic anhydride present in the MAPP not only provides polar interactions, but can covalently link to the hydroxyl groups on the cellulose fibre [10].

PLASMA TREATMENT OF FIBRES

Many wet-chemical, surface treatments for adhesion enhancement are becoming increasingly unacceptable because of environmental and safety consideration. Modification of polymer surfaces by plasma treatment, both corona and low-pressure glow discharges, presents many important advantages.

In the plasma treatment of fibres and polymers, energetic particles and photons generated in the plasma interact strongly with the substrate surface, usually via free radical chemistry. Four major effects on surfaces are normally observed, each of which is always present to some degree, although one may be favoured over the others, depending on the substrate and the gas chemistry, the reactor design, and the operating parameters. The four major effects are surface cleaning, ablation or etching, cross-linking of near-surface molecules and modification of surface-chemical structure.

Plasma treatment can be used with great effect to improve the bond strength to polymer-fibre and polymer-polymer combinations. In these cases, the improved properties result from both increased wettability of the treated substrate on polymeric sheet referred by the adhesive and the modification of surface chemistry of the polymer. The changed surface chemistry facilitates reaction of the adhesive with surface species during curing, to form covalent bonds with the plasma-treated interphase. A detailed critical review on plasma surface modification of polymers for improved adhesion has been made by Liston et al [11].

Wettability and adhesion

Modified wettability is one of the most apparent results of plasma treatment and the method used for characterizing the modification is to measure the advancing and receding contact angles against specific liquids. Plasma-produced polar groups increase the surface free energy, γ, of the fibre and decrease the contact angle, θ, which usually correlate with better bonding of adhesives, and θ has often been used as an estimate of bonding quality.

Hydrophobic polymeric surfaces display hysteresis of advancing and receding contact angles $(\theta_a - \theta_r)$. The lower value of θ_r may depend on the surface roughness, or of polar surface contaminants, or of both. The increasing number of hydrophilic groups as the result of plasma treatment would result in a decrease of $(\theta_a - \theta_r)$.

Many results published have shown that the plasma treatment conditions necessary to achieve maximum bond strength must be optimized for any given material contribution [12-14]. Also the bonding improvements will depend on the polymer formulation, the type and amount of additives, the adhesive, the cure cycle, the time between plasma treatment and bonding, and of course, the plasma process parameters and the type of plasma gas.

Most commercially-available plasma systems are designed for batch operation. There is a great demand for commercially viable plasma systems able to perform continuous, on-line treatment of fabrics and films for coating/laminating applications. Although some "air-to-air" or "cassette-to cassette", batch-continuous systems have been presented and built for yarn, film and fabric treatment, there are still no satisfactory, cheap and effective on-line systems available [15-18].

Low pressure plasma treatment is essentially a batch process with fabric being treated as it is wound from one batch to another. Technoplasma SA has described their plasma machine KPR-180 [19, 20] and Plasma Ireland has reported about the new plasma unit for on-line atmospheric treatment of textiles and nonwovens as the result of a Brite-Euram project "Plasmatex" [21]. The advantages of industrial plasma treatment over the traditional processing can be summarised below in Table XV [22].

Table XV. Plasma treatment vs. traditional processing for textiles

	Plasma	**Traditional**
Medium	No wet chemistry involved. Treatment by excited gas phase	Water-based
Energy	Electricity - only free electrons heated (<1% of system mass)	Heat - entire system mass temperature raised
Reaction type	Complex; many simultaneous processes	Simple
Reaction locality	Highly surface specific, no effect on bulk properties	Bulk of the material generally affected
Potential for new processes	Great potential, field in state of rapid development	Very low, technology static
Equipment	Experimental, rapid development	Mature, slow evolution
Energy Consumption	Low	High
Water Consumption	Negligible	High
Environmental Pollution	Very low	High

Both low pressure and atmospheric plasma systems are being sold and developed by among others, Europlasma in Belgium and Dow Corning in Ireland. These companies are developing new, powerful, generic and highly manufacturable on-line and batch processes for plasma-based, surface engineering of textile fibres and fabrics, nonwovens, plastic films, cellulose-fibre sheets etc.

CONCLUDING REMARKS

Natural cellulose are selected for price, technical properties, performance and for environmental and agronomical reasons. Process-related criteria in technical applications are exchangeability of materials, processing speed, processing costs and amount of waste production

Drawbacks for some ligno-cellulose fibres in technical textiles applications are inconsistency of quality (within and between batches), mechanical behaviour, moisture related properties, heat stability and durability. The gaps in the research are mainly knowledge of fibre extraction technology, chemical and physical fibre characteristics, possible modification of fibres (bulk and surface), processing techniques of the fibres and bio-composites, and the relation between processing technology and the end product.

On-line, objective testing methods are not available for the determination of relevant fibre characteristics. This obstructs to some extent, the introduction of ligno-cellulose fibres as industrial and technical raw materials. There is need for quick analytical methods for determination of the fibre quality at an early stage in the production chain. This aspect is essential for establishing the product and market potential. For industrial application of agro-fibres in different market outlets, it is vital that these fibres can be produced and supplied with a guarantee of quality.

Application of plant fibres requires adaptation of processing technology and adjustment such as pre-treatments, surface modifications etc. Successful implementation of ligno-cellulose based composites will require adjustment of fibre specification to industrial demands. This includes refining, fibrillation, impregnation, coating, cross-linking, grafting, etc.

New fibre extraction techniques, e.g. steam-explosion and extrusion techniques are promising tools for the production of a uniform quality of fibre raw materials for technical applications.

Opportunities for applications for plant fibre products in building and construction materials are obvious.

There are many different potential markets for composites with polymers reinforced with plant fibres. However, research on the modification of the plant fibres is required to provide the industry with fibres miscellaneous characteristics, so that different demands of composite applications can be met. Modifications of natural fibres regarding moisture sensitivity, chemical resistance, temperature resistance and flammability needs to be carried out. As waste management of the plant fibre composite is expected to be advantageous, it would be of great use to support this arrangement with scientific data.

The problems associated with the use of natural fibres , viz., moisture sensitivity, chemical resistance, heat resistance, flammability etc. have to be dealt with seriously.

Since the waste management of the natural fibre-based technical textiles is an important argument, this aspect has to be supported by scientific data.

Technical problems have also to be solved in extrusion processes related to the compounding of ligno-cellulose fibre with polymeric matrices for injection-moulded composites. An important drawback of the use of these types of fibres is the lower processing temperature permissible due to the possibility of thermal degradation and/or the possibility of emissions that could affect the composite properties. The processing temperatures are thus limited to about 200 °C, although it is possible to use higher temperatures for short periods.

The other drawback is the moisture absorption property of the natural fibres and biopolymer matrices. Moisture absorption can result in swelling of the fibres and concerns about the dimensional/form stability of the bio-composite cannot be ignored. It may be difficult to entirely eliminate the absorption of moisture without using some sort of surface barriers on the composite surface. When the problems of dimensional stability and compatibilty between the natural fibres and matrices are solved this could lead to major new markets for agro-based resources and bio-composites.

Potential markets for nonwovens based on plant fibres are: absorbents (hygienic products, sanitary and medical biodegradable nonwovens/disposables); geotextiles (for soil stabilization, drainage, soil erosion control and stimulation of vegetation growth); fillers (for furniture and car seats) and reinforcement (flexible and hard composites). Research topics include adaptation of suitable processing technologies and methods in order to achieve the required fibre and web characteristics.

Potential markets for long plant fibres are reinforcement of paper pulps in general, upgrading of waste paper pulps, and pulp grades for various niche markets. Fibre crops have potential application in areas, other than fibre products, like biomass production for energy generation, as packaging materials, separation media and for chemical and specialities production. Finally cellulose derivatives find many industrial applications as adhesives, binders, films, emulsifiers, thickeners, stabilizers etc.

REFERENCES

1. P S Mukherjee and K G Satyanarayana, *J. of Material Science*, 1986 4162-4168.
2. H Baumgarti and A Schlarb, 'Nachwachsende Rohstoffe-Perspektiven für die Chemie', Proceeding and Symposium, Frankfurt, May 5-6, 1993.
3. W I Chodyrew, *Melliand Textilberichte II,* 1990 825-830.
4. M Matsui, *Melliand International No. 4,* 1996 178.
5. Correspondence from ITF, Lyon.
6. D J Cole, 'Advances in Fibre Science- Chapter 2', Manchester, The Textile Institute, 1992.
7. A S Herrman, et al, Proceeding *Technitex Conference on Technical Textiles*, TURIN, Nov. 21-23, 1966.
8. R Shishoo, Proc. of *Nordflex Conference*, Tampere, Finland, 10-12 August,1998.
9. J Ohlsson, R Shishoo, B Nyström, Proc.of *Techtextil Symposium*, April 2003.
10. J M Felix, P Gatenholm and H P Schreiber, *Polymer Composites,* 1993 **14** 449
11. E M Liston, L Martinu and M R Wertheimer, *J of Adhesion Science Technology*, 7 1091-1127.
12. E M Liston, *J of Adhesion Science Technology*, 1989 **30** 199-218.
13. E Occhiello, M Morra, G Morini, F Garbassi and D Johnson, *J of Applied Polymer Science,* 1991 **42** 2045-2052.
14. J D Moyer and J P Wightman, *Surface Interface Anal*, 1991 **17** 457-464.
15. W Landman, Model 7500S, GaSonics/International Plasma Corporation, 2730 JunctionAvenue, San Jose, CA 95134-1909, USA, product literature.
16. G Liebel, Model 4002-B, Technics Plasma GmbH, Dieselstrasse 22a, DE-8011 Kirchheim bei Munchen, Germany, product literature.

17. W Landman, Model 8150, GaSonics/ International Plasma Corporation, 2730 Junction Avenue, San Jose, CA 95134-1909, USA, product literature.
18. J T Felts, Model "Flex-1", AIRCO Coatings Technology, P.O. Box 4105, Concord, CA 94524 USA, product literature.
19. E Godau, 'Using Plasma Technology', *Textile Technology International*, 1996
20. E Godau, *Proceedings of 17th IFVTCC Congress*, Vienna, June 5-7, 1996, 177-181.21 T Herbert and E Bourdin, "Index 99", Publ. EDANA, Brussels, 1999
21. R Shishoo, *J.of Coated Fabrics*, 1996 **26** 26-35.

SOME PROPERTIES OF KENAF AND KENAF COMBINED WASTE COMPOSITES

W. Y. Wan Ahmad, J. Salleh, M. F. Yahya, M. I. Abdul Kadir, M. I. Misnon
Faculty of Applied Sciences, Universiti Teknologi MARA, 40450 Shah Alam, Selangor
Darul Ehsan, Malaysia

ABSTRACT

Kenaf is one of the most promising fibers for composite conversion. In this work, kenaf is being combined with and combing waste or sheared polyester waste and polyester resin as matrix. The comparison of properties are based on tensile strength and flexural strength of 100% resin, 100% kenaf and 50/50 kenaf/cotton combing waste 50/50 kenaf/sheared polyester waste in three types of lay-ups of longitudinal, cross-laid and transverse. The highest results of tensile strength are from 100% longitudinal kenaf and 100% cotton combing waste but the highest results of flexural strength are 50/50 cross-laid lay-ups of composites. The tensile and flexural strength properties are also compared with similar properties on glass composites.

INTRODUCTION

Composite is any combination of two or more resources, in any form, and for any use (1). The normal fibers used for composites are kevlar, carbon and glass. As a result of global environment awareness, many workers of composites turn to low cost, biodegradable materials with improved properties (2). Bio-based resources are renewable, widely distributed, inexpensive, moldable, available locally, anisotropic, non-abrasive, versatile, easily available in many forms, biodegradable, compostible and reactive (3,4,5). One of the main areas of research in composite is in combining natural fibers with thermoplastics (6). Kenaf is one of the biodegradable and bio-based composites. Kenaf (Hibiscus cannabinus L.) is herbaceous warm-season annual fiber crop related to cotton, okra and hibiscus, which can be grown under a wide range of weather conditions (7). Kenaf is a minor textile fiber which has been used as cordage, ropes, basket weaving and alike. However, it seems that now kenaf is discovered as 'versatile' plant which can generate many products such as food for animals, has herbal medicinal properties, and composite from different parts of the plant (barks, cores and leaves). Composite is one of the strong points of kenaf. It can be used to fabricate low-performance composites. The combination of kenaf with cotton combing waste and sheared polyester waste from textile factories will be more notable and 'waste saving' as it will form a value added material in a successful conversion.

MATERIALS AND METHODS

Kenaf is supplied by the Malaysian Agriculture Research and Development Institute (MARDI) and is the main component of the composites for this work. Polyester resin is used as the matrix to bind kenaf and cotton combing waste. On the other hand, cotton combing waste was donated by CNLT (M) Sendirian Berhad, Senawang, Negeri Sembilan Malaysia and sheared polyester waste was taken from Ara Borgstena Sendirian Berhad, Banting, Selangor Darul Ehsan, Malaysia. The sheared polyester waste is from dyed circular knitted automotive fabrics.

Sample Preparation

MARDI carried out the retting process on kenaf and the bast fibers were supplied in strand form. The fibers were then manually combed to straighten and parallelize them. The combed fibers were then cut in approximately six inches length and arranged for composite fabrication in three lay-ups of longitudinal, cross-laid and transverse. These lay-ups of composites were for kenaf only because cotton combing waste and sheared polyester waste are arranged in random form. The layers of web form cotton combing waste and the layers of sheared polyester waste were also cut in six inches square dimension to be in the same dimension as kenaf fibers.

The 100% kenaf

The combed and cut fibers were manually laid up to form layers of fibers at a certain weight according to specified proportions. Figure 1 shows the lay-up of 100% kenaf fibers for cross-laid composite fabrication whereas Figure 2 shows the kenaf fibers lay-up for the longitudinal and transverse forms of composite fabrication.

Figure 1: Kenaf Lay-ups for Cross-laid Composite

Figure 2: Kenaf Lay-ups for Longitudinal and Transverse Composite

The 50/50 kenaf combined with other fibers

Kenaf Fibers

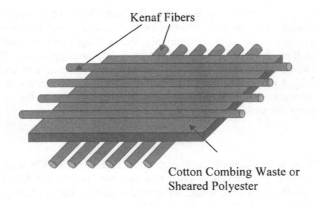

Cotton Combing Waste or
Sheared Polyester

Figure 3: Fibers Lay-ups of 50/50 Combined Kenaf for Cross-laid Composite

The combined samples were prepared by layering cotton combing waste or sheared polyester waste in between layers of kenaf fibers. Figure 3 shows the diagram of the 50/50 kenaf for cross-laid composite. The fiber lay-ups of 50/50 kenaf for longitudinal and transverse composite are shown in Figure 4.

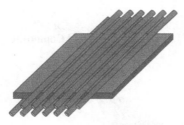

Figure 4: Fiber Lay-ups of 50/50 Combined Kenaf for Longitudinal
and Transverse Composite

The fibers which were already arranged to be converted into composites were then applied with polyester resin and pressed in between two aluminum plates (12" x 12") using G-clamps and 2 mm spacers (thickness controller) for twelve hours (Figure 5).

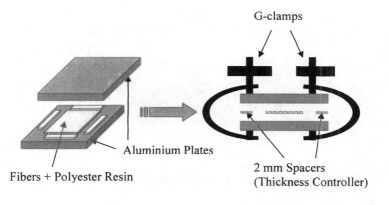

Figure 5: A Typical Composite Fabricating

Mechanical testing

Tensile tests were performed using Testometric Micro – 500 testing machine in accordance with Composite Research Advisory Group, CRAG – 302 standards. The specimens were tested at the rate of 10 mm per minute. Tensile strength was calculated from load-extension curves. The sample dimension is shown in Figure 6.

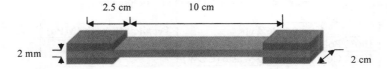

Figure 6: Sample Dimensions for Tensile Test

Flexural rigidity tests were performed on the same testing machine using the three points bending method as per CRAG – 200 standards. The sample dimensions used for this standard testing method are shown in Figure 7.

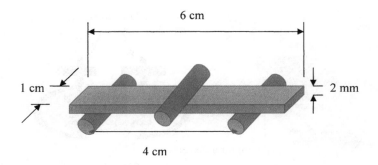

6 cm

1 cm

2 mm

4 cm

Figure 7: Sample Dimensions for Flexural Test

RESULTS AND DISCUSSION

Table 1 and Figure 8 show the value of tensile strength of composites. In terms of tensile strength, the 100% longitudinal kenaf composite has the highest strength of 90.09 Mpa followed by the 50/50 longitudinal kenaf and cotton combing waste composite(75.37 Mpa). The 100% cotton combing composite has a tensile strength of 70.9 MPa whilst longitudinal kenaf and sheared polyester composite has the strength of 50.14 MPa. The other composites lay-ups have lower values than the 100% polyester resin (matrix). Hence, the composites of higher values than the matrix are considered for future fabrication.

Table 1: Tensile Strength of Composites

Composite Types	Tensile Strength (MPa)
100% Polyester Resin	37.73
100% Cotton Combing Waste (CCW)	70.90
100% Sheared Polyester (SP)	27.93
100% Kenaf (Longitudinal)	90.09
100% Kenaf (Cross-laid)	35.45
100% Kenaf (Transverse)	9.56
50/50 Kenaf/CCW (Longitudinal)	75.37
50/50 Kenaf/CCW (Cross-laid)	35.68
50/50 Kenaf/CCW (Transverse)	33.47
50/50 Kenaf/SP (Longitudinal)	50.14
50/50 Kenaf/SP (Cross-laid)	33.78
50/50 Kenaf/SP (Transverse)	19.49

Similar work on composite as kenaf, cotton combing waste and sheared polyester waste has been carried out on glass fiber but using epoxy resin as matrix. The tensile results are shown in the Table 2. In all cases, the tensile strength of glass fabric is much higher than tensile strength of 100% longitudinal kenaf. The work on polypropylene matrix with non-woven glass composite gave a tensile strength of 100 Mpa (8).

132

Table 2: Tensile Strength of Glass Composites

Type of fabric	Layers of fabric	Strength (MPa)
Plain weave glass	12 layers (2 mm composite)	149.7 (Warp) 160.9 (Weft)
Satin weave glass	10 layers (2 mm composite	268.7 (Warp) 208.2 (Weft)

Tensile Strength of Composites

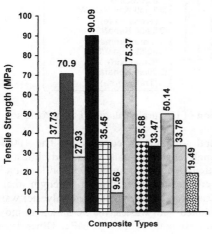
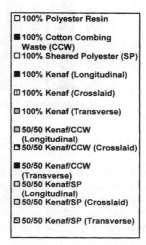

Figure 8: Graphical Presentation of Tensile Strength of Composites

Table 3: Flexural Strength of Composites

Composite Types	Flexural Strength (MPa)
100% Polyester Resin	86.20
100% Cotton Combing Waste (CCW)	126.85
100% Sheared Polyester (SP)	96.79
100% Kenaf (Longitudinal)	125.98
100% Kenaf (Cross-laid)	65.71
100% Kenaf (Transverse)	20.29
50/50 Kenaf/CCW (Longitudinal)	133.07
50/50 Kenaf/CCW (Cross-laid)	136.29
50/50 Kenaf/CCW (Transverse)	35.78
50/50 Kenaf/SP (Longitudinal)	139.50
50/50 Kenaf/SP (Cross-laid)	149.62
50/50 Kenaf/SP (Transverse)	52.00

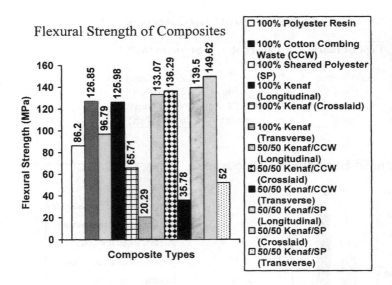

Figure 9: Graphical Presentation of Flexural Strength of Composites

Table 3 and Figure 9 show the flexural strengths of composites. The 100% cotton combing waste and the 100% longitudinal kenaf composites recorded almost the same flexural strength. However, the higher strength comes from 50/50 kenaf longitudinal and cross-laid. In fact 50/50 cross-laid composites have higher strength than 50/50 longitudinal composites for both cotton combing waste and sheared polyester waste. And comparing both of the 50/50 kenaf combined with wastes, the 50/50 kenaf and sheared polyester waste has the highest flexural strength of 149.52 MPa. Other lay-ups for composites show lower flexural strength than the matrix. Comparison with glass composites as in Table 4 indicates that the kenaf combined with cotton combing waste and sheared polyester waste composites has better flexural strength than glass composites.

Table 4: Flexural Strength of Glass Composites

Type of fabric	Layers of fabric	Strength (MPa)
Plain weave glass	12 layers (2 mm composite)	68.8 (Warp) 63.2 (Weft)
Satin weave glass	10 layers (2 mm composite	85.8 (Warp) 70.1 (Weft)

CONCLUSION

It has been indicated that natural fiber materials have the potential to compete with glass fiber in composite materials (9-14). This has been clearly shown in the flexural strength property of kenaf and combined waste composites. The result of kenaf and combined wastes will be higher if the preparation and parallelization of kenaf fiber lay-ups can be improved using machines that can straighten and parallelize the kenaf fibers. It could be at par with glass composites and at the same time the flexural strength may be even much higher than glass composites.

ACKNOWLEDGEMENT

The authors wish to thank MARDI, CNLT (M) Sdn Bhd, Ara Borgstena Sdn Bhd for their cooperation in supplying raw materials and M. K. Yusoh who performed part of the testing.

REFERENCES

1. B. English, P. Chow, and D.S. Bajwa in Paper and Composites from Agro-Based Resources, R.M. Rowel, R.A. Young, J.K. Rowell, Editors, (1996), CRC, New York, 270
2. G. Canché-Escamilla, J. Rodriguez-Lavadia. J.I. Cauich-Cupul, E. Mendizbal, J.E. Puig and P.J. Herrera-Franco, Composites: Part A **33** (2002) 539-549
3. R.M. Rowel, Proceedings, The Fourth Pacific Rim Bio-based Symposium, November 2-5, (1998), Indonesia, 1
4. P. Hedenberg, and P. Gatenholm, Conversion of Plastic Cellulose Waste into Composites 1. Model of the Interphase. J. Applied Polymer Science (1995), **56**, 6, 641-651
5. V. Favier, H. Chanzy and J.Y. Cavaille, Polymer nanocomposites reinforced by cellulose whiskers, Macromolecules (1995) **28**, 18, 6365-6367
6. A.R. Sanadi, D.F.Caulfield and Rowell, R.M., Reinforcing polypropylene with natural fibers. Plastic Engineering, Vol. L, 4, 27
7. J. Johnson, (April 2001). What is Kenaf. Rural Enterprise and Alternative Development Initiative Report. Retrieved July 5th, 2003 from http://www.siu.edu/~readi/grains/factsheets/kenaf.pdf
8. H. Baumgartl and A. Schlarb, 2. Symposium Nachwachsend Rohstoffe-perspektiven für die Chemie, Frankfurt, 5-6 May 1993
9. T. Sterzynski, B. Triki and S. Zelzany, Polimery **40**, 7/8 (1995) 468
10. R.G. Raj and B.V. Kokta, Proc Ann Tech Conf ANTEC' 91, SPE (1991) 1883
11. K.P. Mleck, A. Nechwatal and C. Knobelsdorf, Die Angew Makromol Chem (1995) 225 73
12. K.P. Mleck, A. Nechwatal and C. Knobelsdorf, Die Angew Makromol Chem (1995) 225 37
13. P.R. Hornsby, E. Hinrichsen, K. Taverdi and J. Mater, Sci (1997) **32** 1009
14. N.E. Zafeiropoulos, D.R. Williams, C.A. Baillie and F.L. Matthews, Composites: Part A 33 (2002) 1083-1093

TENSILE AND FLEXURAL PROPERTIES OF COMPOSITES MADE FROM SPINNING WASTE

J. Salleh, W.Y.W. Ahmad, M.R. Ahmad, M.F. Yahya, S.A. Ghani and M.I. Misnon
Faculty of Applied Sciences, Universiti Teknologi MARA, Malaysia

ABSTRACT

Spinning especially combed yarn produces considerable waste fibre. These fibres are sold at very low price for fillings. An investigation of converting these wastes to some form of value-added material was successful in forming a composite material made from natural fibre.

Cotton combing noils and blowing waste were fabricated into thin composite boards using polyester resin at room temperature utilising a compression method. Tests on tensile and flexure properties of these composites were evaluated against 100% polyester resin plaques. It was found that composites made from cotton waste were stronger than polyester without the reinforcement. Some possible applications of these composites are thin boards or panels that can be used to replace wood and fibre-board products.

INTRODUCTION

Cotton is the most widely used natural fibre in the world. The demand for cotton has always been strong and the trends over the years, suggesting cotton is holding its own market share while polyester has grown to be by far the most popular man-made fibre. The processing of fibres to yarn and fabric produces considerable waste. The practice of recovering waste is perhaps as old as the art of spinning and weaving. In cotton spinning, short fibres are collected as waste from blowroom to combing. Waste cotton fabrics or even worn-out garments can be garnetted to produce very short fibre. This waste cotton is too short a fibre to be useful for textile applications except as filling material and for cleaning cloths. However, it can be converted into some value-added products such as composites.

Traditional high performance fibre-reinforced plastics use carbon, aramid and glass fibres, as reinforcing fillers. They dominate the aerospace, leisure, automotive, construction and sporting industries. Glass fibres are the most widely used to reinforce plastics due to their low cost (compared to aramid and carbon) and fairly good mechanical properties. However, these fibres have serious drawbacks such as high cost, are non-recyclable, and non-biodegradable, pose high health risks when inhaled, high high energy consumption and high density.

Recently there has been a growing trend of making completely new types of composites by combining different resources using natural fibres as the base fibre. A bio-based composite can be defined as any combination of two or more resources held together by some types of mastic or matrix system[1]. Normally for a composite, the fibres will act as reinforcing filler material and the matrix will serve as a binder and stiffen the resulting assembly. The objective is to combine two or more materials in such a way that synergy is created between these components and thus will result in a new material that is better than any of the individual component.

Recent global environmental issues and awareness in preservation of natural resources and the need for recycling has led to a renewed interest in using natural fibres for composites. Natural fibres are also in general suitable for use in reinforcing low

136

performance composites. They are claimed to exhibit many advantageous properties such as having low-density material, and yielding relatively lightweight composites with good specific properties and are non-brittle (shatter resistant). Natural fibers are also a highly renewable resource which reduces the dependency on foreign and domestic petroleum oil. There may also be significant socio-economic benefits in terms of rural jobs generation and enhancement of the non-food agriculture economy[2,3]. However, the overall physical properties of those composites are far away from the performance of glass-fibre reinforced composites.

Bledzki and Gassan[4] reported that natural fibers were used as early as 1908 in the fabrication of large quantities of sheets, where paper or cotton was used to reinforce sheets made of phenol-or melamine-formaldehyde resins. Natural fibres like flax, sisal, jute, coir and kenaf have all been proved to be good reinforcement materials in thermoset and thermoplastic matrices[5-13]. Many applications of natural fibre composites have already been established in low performance materials. Some common examples are panels for automotive and buildings, pipes and packaging materials[4]. However, many challenges exist in identifying market segments, determining performance requirements, selecting resins, and optimizing manufacturing properties.

Many papers have been published about the effect of chemical treatments on composite properties[2,7,11-14]. Typical and relatively simple treatments, such as an alkali and silane treatments, have proved their usefulness. Work on thermoplastic matrices such as polypropylene and polyethylene[7-8,13,14] showed improvement in composite properties. However, work using thermosets[15-18], especially the epoxides have shown some contradictory findings regarding the influence of coupling agents. More studies on suitable coupling agents are needed.

Waste fibres from textile processing activities can be a source of reinforcement filler for composites. Even in these days of increased environmental awareness, millions of tonnes of waste products from textiles are landfilled each year. Therefore the abundance of these fibers has created the need to develop alternative recycling avenues. One such alternative is their possible conversion to some value-added products. In this project, the suitability of cotton waste as a raw material for the production of composite panel products has been determined. Cotton waste from spinning mills (blowing and combing waste) was processed to form composites using polyester as matrix. Tensile and flexural properties of the composites were assessed.

EXPERIMENTAL

Materials

Blowing wastes from a cotton spinning mill were processed into a continuous fleece of fibres using a scutcher. The sheet of fibres was separated by inserting paper into each layer. Combing noils were also collected from a Cherry-Hara combing machine. At first the noils were randomly packed together and it was found that it caused uneven layers of fibre web during lay-up. Subsequently, the noils were collected in a more systematic manner where each layer was separated by paper. The types of fibre-reinforced composites fabricated from cotton waste were: 100% comb noils (PCN), comb noils + silane treatment (PCN-S), 100% blowroom waste (PBW), blowroom waste + comb noils 1:1 (PCNBW).

Composites manufacture

Waste fibre fleeces of 150 mm x 150 mm were evenly arranged on an aluminum plate measuring 300 mm x 300 mm. Several trials were conducted to find the suitable number of layers needed to fabricate into a 2 mm thick composites. These laminates were then pressed between a pair of aluminum plates and were left under pressure for 24 hours in order to pack the fibres together.

Composite sheets were manufactured by impregnating each fibre laminate with polyester resin and 2% methyl ethyl ketone hardener. A 40:60 ratio of fibre to resin was prepared. Air bubbles were removed carefully by squeezing with a steel roller. The impregnated laminate was then compressed under a pair of aluminum plates as shown in Figure 1. A spacer of 2 mm thickness was inserted between the plates. The assembly was left for 24 hours to cure.

Chemical treatment of fibres using a silane agent was undertaken by wetting the combing noils with a 5% w/w silane solution. Excess solution was squeezed using a padding mangle and the treated sheet was dried in an oven at 80°C. The dried comb noils laminates were then processed into composites in the same manner above. For comparison purposes, a 100% polyester plate of similar dimension was also made using a mould.

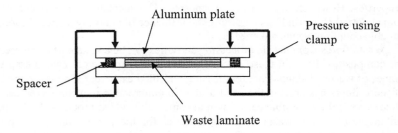

Figure 1: The Assembly for Composite Manufacture

Test methods

Tensile tests were conducted using a Testometric 500 tensile tester according to the CRAG (Composite Research Advisory Group) test method 302. Each end of the specimen was tabbed with the same material using epoxy adhesive to avoid damage at the grips. The width and gauge length were 20 mm and 100 mm respectively and all tests were carried out at a crosshead speed of 10 mm min⁻¹.

Flexural tests were conducted on the same machine using a three-point bending test in accordance to the CRAG 200 test method. The span to thickness ratio should be high enough to produce bending failures. The top and bottom rollers were 10 mm in diameter and the span length was 40 mm. The specimen dimension was 60 x 10 x 2 mm and a cross-head speed of 15 mm min⁻¹ was used.

RESULTS AND DISCUSSION

Tensile tests

Figure 2 shows a comparison of the measured tensile strength of cotton waste composites. Composites made from combed noils displayed the highest tensile strength (71 MPa) while the polyester resin plate without any fibre reinforcement showed the lowest (38 MPa). In fact, all composites with fibre reinforcement showed higher values than polyester plate. It is evident that cotton waste reinforcement contributes to the tensile strength of composite materials. On average, the fibre volume fraction for waste cotton composites was about 52% except comb noils + blowing waste which recorded only 46%.

It is also apparent from Figure 2 that the longer the waste fibre, the higher will be the tensile strength. Comb noils composites, which consist of longer fibres than blowing waste, show higher strength. However, the experiment with silane shows that the strength is weaker. Although silane is said to improve the properties of glass/epoxy and bio-based thermoplastic composites, it seems that this is not the case for polyester resinated composites. As mentioned earlier, works using epoxides have also shown some contradictory findings; hence more studies are needed here.

Breaking strain values which are shown in Figure 3 also exhibit similar results as tensile strengths where comb noils-containing composites record the highest value (up to 8%). However, the strain readings were taken directly from tensile machine print outs without the use of strain gauges, hence the data may not be reliable. Nevertheless, the values maybe used for comparison purposes of this experiment. Thermoset polyester plate is brittle and breaks at only 2.7%. Again it seems that the longer fibres of comb noils could have held together longer before they fail.

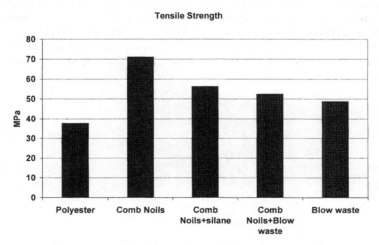

Figure 2: Tensile Strength - Waste Cotton Fibre-reinforced Polyester

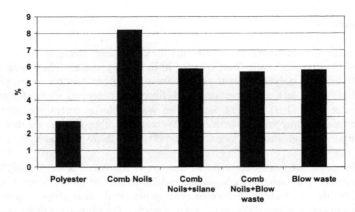

Figure 3: Breaking Strain - Waste Cotton Fibre-reinforced Polyester

Flexural test

Figure 4 shows the result of three-point bend test for cotton waste composites. It should be noted that the flexural test is primarily for material control purposes and does not provide reliable tension and compression data. The composite made from combed noils again displayed superior properties and had the highest flexural strength of 126.4 MPa. Blowing waste composite showed the lowest (83.6 MPa), but the polyester plate also shows somewhat similar results to the blowing waste. Although the mechanisms of failure for tensile and flexural test are different, the trend seems to be similar.

Figure 5 shows the flexural deflection of the composites. The failure mechanism of the fibre-reinforced composites showed the importance of fibre reinforcement of these composites as none of them failed catastrophically. The flexural deflection of untreated cotton waste is almost similar; however the silane-treated composite shows a lower reading and only a little above the polyester plate value. Again, this needs further investigation.

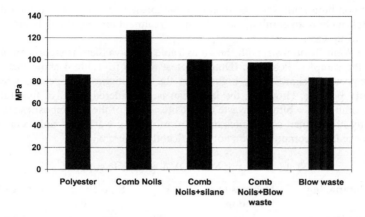

Figure 4: Flexural Strength - Waste Cotton Fibre-reinforced Polyester

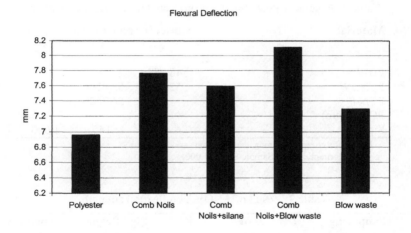

Figure 5: Flexural Deflection - Waste Cotton Fibre-reinforced Polyester

Statistical analysis

An analysis of variance using SPSS 11.5 for Windows was conducted to determine whether any significant difference existed between the treatments. All pairwise multiple comparisons were also made following Post-Hoc test (Duncan). Tables 1 and 2 show the results of these statistical analyses for tensile strength.

From SPSS results, it can be said that comb noils composite is significantly better than all other composites and 100% polyester plate is significantly lower than the others. However, within the experimental error, it can be said that the tensile strength of comb noils + blowing waste and 100% blowing waste composites can be considered equal. It can also be said that the tensile strength of combing noils + blowing waste (1:1) and combing noils + silane treatment composites are similar. On the other hand, composites made from combing noils + silane treatment are significantly better than the 100% blowing waste.

ANOVA on flexural strength shown in Table 3 shows there is significant difference between the samples. Post Hoc (Duncan) test results in Table 4 show that the comb noils composite is significantly higher than the next highest composites which is the comb noils + silane. However, the latter can be considered equal in flexural strength with the comb noils + blowing waste composites. Both of these composites are however statistically higher than composites made from blowing waste and polyester plate, and both of these latter composites are not significantly different.

Table 1: Result of ANOVA on tensile strength.

Comparing	Degrees of Freedom	F	Significance at 0.01	Probability
All Materials	5,20	46.369	Significant	0.000

Table 2: Results of Post-Hoc (Duncan) tests for pairwise comparison

Material	N	Subset for alpha = .05			
		1	2	3	4
P	5	37.5920			
PBW	5		48.7340		
PCNBW	5		52.4260	52.4260	
PCN-S	5			56.2520	
PCN (2mm)	5				71.0060
Sig.		1.000	.159	.145	1.000

Means for groups in homogeneous subsets are displayed.
a Uses Harmonic Mean Sample Size = 5.000.

Table 3: Result of ANOVA on flexural strength.

Comparing	Degrees of Freedom	F	Significance at 0.01	Probability
All Materials	5,20	27.851	Significant	0.000

CONCLUSIONS

Waste cotton from spinning process can be converted into useful products such as composites. The tensile and flexural properties of composites made from these cotton spinning wastes were studied. It seems that the tensile and flexural strengths of the fibre-reinforced composites increase with length of fibres. Among all the composites tested, those made from comb noils recorded the highest values of both tensile and flexural strength and the polyester plate without fibre reinforcement, the lowest.

Table 4: Results of Post-Hoc for pairwise comparison using Duncan (flexural strength).

Material	N	Subset for alpha = .05		
		1	2	3
PBW	5	83.6300		
P	5	86.1980		
PCNBW	5		97.4060	
PCN-S	5		100.088	
PCN (2mm)	5			126.848
Sig.		.583	.566	1.000

Means for groups in homogeneous subsets are displayed.
a Uses Harmonic Mean Sample Size = 5.000.

For tensile strength, the order of strongest to weakest composites can be written as follows:
Comb noils > comb noils + silane > comb noils + blowing waste > blowing waste > polyester plate.
The order of flexural strength, strongest to weakest, is as follows:
Comb noils > comb noils + silane, comb noils + blowing waste > blowing waste, polyester plate.

ACKNOWLEDGEMENT

The authors would like to thank IRDC (UiTM) for sponsoring the project, Chempaka Negri Laksmi Spinning Mill for supplying the waste fibres.

REFERENCES

1 R M Rowell, 'The state of art and future development of bio-based composite science and technology towards the 21st century', Proceedings, the fourth *Pacific Rim bio-based composites* symposium, November 2-5, Indonesia. p.1-18, 1998.

2 I V D Weyenberga, J Ivensa, A D Costerb, B Kinob, E Baetensb, I Verpoesta, 'Influence of processing and chemical treatment of flax fibres on their composites', *Composites Science and Technology*, 2003 **63** 1241–1246.

3 R M Rowell, A Sanadi, R Jacobson and D Caulfield, Ag & Bio Engneering, 1999, Chapter 32, p. 382.

4 A K Bledzki, J Gassan, 'Composites reinforced with cellulose based fibres', *Progress in Polymer Science*, 1999 **24** 221–74.

5 M Hautala, A Pasila and J Pirila, 'Use of hemp and flax in composite manufacture: a search for new production methods', *Composites: Part A*, 2004 **35** 11-16.

6 P Wambua, J Ivens and I Verpoest, 'Natural fibres: can they replace glass in fibre reinforced plastics?', *Composites Science and Technology*, 2003 **63** 1259–1264.

7 K Joseph, S Thomas and C Pavithran, 'Effect of chemical treatment on the tensile properties of short sisal fiber-reinforced polyethylene composites', *Polymer*, 1996 **37** 5139–45.

8   M A Dweib, B   Hu, A O'Donnell, H W Shenton and R P Wool, 'All natural composite sandwich beams for structural applications', *Composite Structures*, 2004 **63** 147-157.

9   K L Fung, X S Xing, R K Y Li, S C Tjong and Y -W Mai, 'An investigation on the processing of sisal fibre reinforced polypropylene composites', *Composites Science and Technology*, 2003 **63** 1255–1258.

10 I K Varma, S R Ananthakrishnan and S Krishnamoorthi, 'Comp of glass/modified jute fabric and unsaturated polyester', *Composites*, 1989 **20** 383.

11 V G Geethamma, M K Thomas, R Lakshminarayanan and S Thomas, Composite of short coir fibers and natural rubber: effect of chemical modification, loading and orientation of fiber.polymer, 1998 **39** 1483.

12 J Gassan and A K Bledzki, 'Possibilities of improving the mechanical properties of jute/epoxy composites by alkali treatment of fibers', *Composite Science and Technology*, 1999 **59** 1303–9.

13 A K   Rana, A   Mandal and S   Bandyopadhyay, 'Short jute fiber reinforced polypropylene composites: effect of compatibiliser, impact modifier and fiber loading', *Composites Science and Technology*, 2003 **63** 801–806.

14   P J Herrera-Franco* and A Valadez-Gonza´lez, 'Mechanical properties of continuous natural fibre-reinforced polymer composites', *Composites: Part A*, 2004 **35** 339–345.

15 S D Varma, M Varma, I K Varma., *J Reinf. Plastic Composites*, 1985 **19** 419-431.

16 E T N Bisanda, M P Ansell, *Composite Science Technology*, 1991 165-178.

17 J Gassan, A K Bledzki, 6th Int. Techtextil Sym. Frankfurt, 1994.

18 J Gassan, A K Bledzki, *Die Angew Makromol Chem*, 1996 **236** 128-138.

UK TECHNICAL TEXTILES: ISSUES RELATING TO SUSTAINABILITY

Brian McCarthy[1] and Chris Byrne[2]

[1]TechniTex Faraday Partnership, BTTG, Wira House, West Park Ring Road, Leeds, LS16 6QL, UK
[2]Mediatex, Ivy Cottage, Park Road, Combe, Witney, Oxfordshire, OX29 8NA, UK

ABSTRACT

This paper reviews recent trends in the UK Technical Textiles manufacturing sector and sub-sectors over recent years. It reviews the activities of the TechniTex Faraday Partnership during the first three-year period of operation (2000 to 2003). It will give examples of the successful transfer of technology from the UK science base into the marketplace. It will focus on establishing the R&D needs of six key sectors within technical textiles and will address a series of issues relating to sustainability.

INTRODUCTION

The textiles industry is a global industry. Across the world, companies are seeking to expand and grow based on product and process innovation. Increasingly, there are demands to 'get science out of the lab and onto the balance sheet'(1). This issue is being recognised in all economies. 'The Indian Textile sector has to be technology driven to retain its position as a major player in the global textile trade' (2).

It is apparent that compared with other countries, British business is not research intensive, and its record of investment in R&D in recent years has been unimpressive. UK business research is concentrated in a narrow range of industrial sectors (e.g. pharmaceuticals, biotech, aerospace, etc), and in a small number of large companies (e.g. GlaxoSmithKline, BAE Systems, Rolls-Royce, BT). All this helps to explain the productivity gap between the UK and other comparable economies. However, there are reasons to be optimistic. Britain's relatively strong and stable economic performance in recent years will improve the climate for business investment of all kinds. Public spending on science is increasing significantly in real terms, and the UK's science base remains strong by international standards, whether measured by the quality or the productivity of its output. The UK R&D tax credit provides an important new incentive for business investment (1).

In addition, there has been a marked culture change in the UK's universities over the past decade with increasing attention paid to spin-out companies. Most of them are actively seeking to play a broader role in the regional and national economy. The quality of their research in science and technology continues to compare well against most international benchmarks. The UK has 90 Universities, 115 Institutions and 56 Colleges of Higher Education with a combined income of £13.5 billion.

THE UK TECHNICAL TEXTILES SECTOR

Business is changing too. Growing numbers of science-based companies are developing across the country, often clustered around a university base. New networks are being created to bring business people and academics together, often for the first time. The UK has real strengths in the creative industries, which are also learning to co-operate with university departments of all kinds (3). However, it is claimed that only 16% of UK businesses engage with the science base.

The UK technical textiles sector numbers some 350 manufacturing companies directly manufacturing fibres, yarns and fabrics with over 700 companies in the broader supply chain. Other textile manufacturers produce some element of technical textiles in their overall output.

FARADAY PARTNERSHIPS

Faraday Partnerships, a flagship UK government initiative, aim to encourage businesses to work with the science base. These alliances include businesses, universities, Research and Technology Organisations, professional institutions and trade associations. There are now 24 partnerships involving over 60 university departments, 27 independent research organisations, 25 intermediary organisations and more than 2,000 businesses – large and small. The core activities of the partnerships include the two-way exchange of information between business and universities, collaborative R&D and development projects, technological and dissemination events. Each Faraday receives £1 million from the UK Research Councils to establish core research activities in their own field or sector. The Lambert Review concluded that Faraday Partnerships can play a valuable role as an intermediary between business and universities.

A Faraday Partnership promotes improved interaction between the UK science, engineering and technology base and industry. A Faraday Partnership is an alliance of organisations and institutions, which can include Research and Technology Organisations, universities, professional institutions, trade associations and firms, dedicated to the improvement of the competitiveness of UK Industry through the research, development, transfer and exploitation of new and improved science and technology.

Faraday Partnerships (2) are dedicated to improving the competitiveness of UK industry through more effective interaction between the science and technology base and industry. Effective interaction requires the identification of industry needs and the subsequent synthesis of the knowledge and experience of those who can satisfy these needs. Crucially, each Faraday Partnership employs a number of technology translators - people with broad experience of knowledge transfer - who can facilitate projects between Partnership members. Established Faraday Partnerships are widely recognised for their technological expertise and understanding of industry's needs. Industry therefore benefits from interactions with the relevant Faraday Partnership(s) and participation in their activities when embarking on new product and process development.

Faraday Partnerships aim to:

- be widely recognised for their technical expertise and be UK industry's first choice for help with new product and process development.
- provide better ways of exploiting R&D to create new products and processes and provide more effective and coherent uptake of the various support mechanisms available (and provide of human and financial resources) e.g. TCS/KTP, LINK, CASE awards, SMART, International Technology Service, Eureka, European Union Framework Programmes.
- link many different organisations, each with a part to play in delivering the Partnership objectives.

- deliver the four 'Faraday Principles'.

The Faraday Principles:

1. Promoting active flows of people, science, industrial technology and innovative business concepts to and from the science & engineering base and industry.
2. Promoting the partnership ethic in industrially relevant research organisations, business and the innovation knowledge base.
3. Promoting core research that will underpin business opportunities.
4. Promoting business-relevant post-graduate training, leading to life-long learning.

TechniTex is a partnership between BTTG, Heriot-Watt University, the University of Leeds and UMIST (3). The clear focus is the UK Technical Textiles sector. The Chairman of TechniTex is Lord Simon Haskel – Past World President of the Textile Institute and Deputy Speaker of the House of Lords.

The UK technical textiles industry sector and market has always been difficult to define, especially for the purpose of collecting useful and consistent statistics.

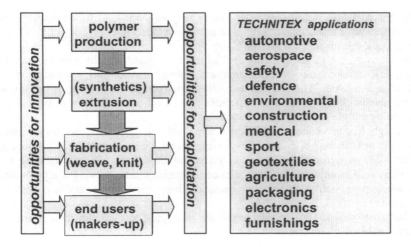

Textile fibres, yarns and fabrics are used in a wide range of downstream applications such as reinforcement, protection, insulation, absorbency, filtration etc. and there are continual definitional problems about exactly where the boundaries of the textile industry and its activities lie. For example, the coating of PVC or rubber onto a textile fabric is, by convention, considered to be a textile process whereas the manufacture of a tyre, also involving the bonding of textile reinforcement to rubber, is normally not so regarded. Numerous other 'grey' areas exist. Many fabricators and processors of technical textiles (and, indeed, some manufacturers of new-generation textile materials such as nonwovens) are reluctant to recognise themselves as part of 'the textile industry' at all.

Problems also arise with regard to the increasing overlap between technical textiles and traditional consumer applications (clothing, interior textiles etc.). Protective clothing technologies and functions have diffused into many areas of performance, sport and leisure clothing while most furnishing, household and decorative textiles now have strong technical functions, especially fire retardency (4).

Overall demand for technical textiles in the UK appears to have peaked at around £1.5 billion in 1996 after a number of years of strong growth in the mid-1990s. Since then, and even more markedly since 2000, there appears to have been an accelerating decline in UK total output and consumption, primarily due to be a long term movement of downstream manufacturing operations to off-shore locations. Total usage of technical and industrial textiles in the UK had fallen to around £1.2 billion by the year 2002. However, from 1998 to 2004 exports rose by 12%.

However, these bare statistics conceal some very different trends, with continuing decline in some older, more traditional areas of 'industrial' textiles but exciting growth possibilities in newer, more technical products and application areas.

Mature product areas particularly under threat include:

- sewing thread
- 'canvas goods', including tarpaulins, sacks, bags etc
- tyre cord
- waddings, cotton wool etc.
- coated fabrics

The coated fabric industry appears particularly vulnerable to future decline, even though coaters have made valiant efforts to increase exports over recent years. There is very little slack left for them to take up.

Undoubtedly, the fastest growing area of technical textiles is the nonwovens sector although, UK nonwovens manufacturers are failing to take full advantage of the opportunities available. The UK has undoubted strengths at the heavier end of market, including needlefelt and stitchbonded fabrics but in other areas there is high import penetration by a highly globalised industry.

Although trends in the coated fabric and made-up goods sectors may have an increasing knock-on impact on the UK weaving sector, there are still some encouraging signs here. The weaving of high tenacity yarns and glass yarns are both showing signs of growth as well as enjoying healthy balances of trade. Glass weaving is a particularly strong area, presumably driven by demand from the composites and asbestos replacement markets.

Other areas where UK manufacturers are managing to maintain their position in reasonably stable markets are:

- mechanical rubber goods
- rope, twine and netting
- papermaking felts.

However, UK nonwovens manufacturers are failing to take advantage of opportunities in this rapidly growing market. Other than some UK strengths at the heavier end of nonwovens, including needlefelt and stitchbonded fabrics, the outlook is one of increasing import penetration by a highly globalised industry.

Overall, the UK has a net trade deficit of about £0.15 (12% of net supply) in technical textiles. However, this disguises the fact that the UK exports some 68% of its output and imports 67% of its requirements. The technical textile market is a highly segmented one with large volumes of specialised products being traded in both directions.

EXAMPLES OF SUCCESSFUL INNOVATION AND EXPLOITATION

The Cancer test bra – being developed by Dr Wei Wang (Biomedical Engineering Unit, De MontFort University, UK). The bra works by detecting abnormal breast cells with electrical currents – a non-invasive process. The bra is being tested by the Tianjin Virtual Engineering Co. in China.

The introduction of flight socks for the prevention of embolism during long haul flights by SSL plc of Oldham, UK – (4)

Electrospinning of carbon nanotubes – Prof. Alan Windle – University of Cambridge, UK

Praybourne Products, UK have incorporated electro-luminescent cables into PPE workwear to provide active protection.

Laser technology applied to textiles – Dr Martin Sharp – University of Liverpool, UK

Novel medical textile implants – Mr Julian Ellis – Ellis Developments, UK

Each emerging technology has associated sustainability issues.

SUSTAINABILITY ISSUES AFFECTING THE SECTOR

As high-volume traditional textile manufacturing continues to move towards China and the Pacific Rim, developed countries – Japan, Koreas, Germany, Italy, etc. – are investigating and investing in the technical textiles sector.

In traditional textiles proactive strategies have been adopted by the textiles sector based on eco-efficiency, cleaner technologies and product stewardship. For example, UK-based Interface Fabrics have adopted the following approach:

1. Eliminate Waste - The first step to sustainability: eliminating waste and the concept of waste, not just incrementally reducing it.
2. Benign Emissions - Eliminating molecular waste emitted with negative or toxic impact on our natural systems.
3. Renewable Energy - Reducing the energy used by our processes while replacing non-renewable sources with sustainable ones.
4. Closing the Loop - Redesigning our processes and products to create cyclical material flows.
5. Resource Efficient Transportation - Exploring methods to reduce the movement of molecules (products and people) in favour of moving information via plant location, logistics, information technology, video-conferencing, e-mail and telecommuting.
6. Sensitivity Hook-up - Creating a community within and around Interface that understands the functioning of natural systems and our impact on them.
7. Redesign Commerce - Redefining commerce to focus on the delivery of service and value instead of the delivery of material; engaging external organisations to create policies and market incentives that encourage sustainable practices.

The sector will be faced with future drivers and trends including (5):

- New multi-fibre agreement in 2005
- SA8000 and social accountability
- EMAS regulation and supply chain management
- Green purchasing networks
- Retailing supply chain management

- Financial rating and institutional investors
- GRI reporting initiative
- Indicators and benchmarking networks
- Consumer boycott and sportswear

The UK textile banks in particular – charity clothes to developing countries – have serious issues with low value waste and are heavily reliant on landfill.

Technical textiles raise particular issues:

- The fibre is selected to provide performance rather than aesthetics – it may be over-engineered to provide guaranteed functionality
- The technical textile may be a sub-component of a larger assembled unit with potential recycling implications
- Performance and functional fabrics are increasingly made of fibre blends – again with recovery and recycling implications
- At the lower end of technical textiles (e.g. sacks, etc.) with cheaper material used in reasonably high volume – there is a need for an environmental impact assessment
- At the higher end some fibres produce great improvements in functionality and performance in use that far outweigh the associated environmental impact

The sector needs to respond to these challenges by greater investment in R&D and innovation. Two areas of particular interest have been identified. The first relates to increasing the service life of products (e.g. personal protective equipment for fire-fighters) by providing novel in service after-care chemical treatments. The second relates to 'smart' tagging of the textile product or sub-component to aid the tracking of the material during its working life and to assist recycling or reprocessing.

REFERENCES

1. The Lambert Review, DTI, 2003
2. www.faradaypartnerships.org
3. www.technitex.org
4. www.schollflightsocks.com
5. www.inem.org

Part IV

Waste management

THE EFFECT OF A FLOCCULENT ON THE COLOUR REMOVAL PROPERTIES OF A CONDITIONED ACTIVATED SLUDGE

Gill Smart[1] and John Binkley[2]

[1]Centre for Materials Research and Innovation, Bolton Institute, Deane Road, Bolton, BL3 5AB, UK
[2] Foundation Studies, Maths and Social Sciences Building, University of Manchester, PO Box 88, Manchester, M60 1QD, UK

ABSTRACT

Colour in effluent is a problem for dyers and many small dyers need to find an easy and economic solution to the problem. Space is often at a premium and small dyers are not large enough to justify the expense of an on-site treatment plant and the local water authority will not accept coloured effluent. One of the easier solutions is to introduce a commercial flocculent into the effluent as it enters a large tank thereby incorporating the colour into flocs, which can be collected and disposed of separately. This investigation finds the optimum flocculent concentration for total colour removal, thus economizing on flocculent usage. The superior colour removal properties of conditioned, activated sludge are shown not to be the result of any residual flocculent remaining within the sludge flocs. The flocculent was added to a laboratory scale activated sludge system using a municipal activated sludge from the local sewage treatment plant and greatly improved the colour removal capabilities of the sludge.

INTRODUCTION

Industrial processes have always had a great impact on the natural environment with liquid effluent being discharged to sewer or water course. A clean-up campaign, initially by the UK National Rivers Authority (NRA) and latterly by the Environment Agency (EA) over recent years has meant that any coloured effluent discharged into a watercourse has become more noticeable. This discharge can come from the local sewage treatment works (STW) or an on-site industrial treatment plant with a discharge permit. Colour was not considered to be a problem in the past but as waterways have become cleaner, there has been an increase in their recreational use with a corresponding concern over colour (Pierce, 1994). Colour in water is visible by the naked eye at very low concentrations and also has a detrimental effect on the transmittance of light with a subsequent deterioration in the aquatic flora and fauna (Diaper et al, 1996).

Approximately 36% of worldwide fibre production is cotton with these figures set to remain for the foreseeable future (Anon, 2002) and much of this will be dyed with reactive dyestuffs to give the textiles the bright colours and high wash and light fastness that the consumer demands. Treating textile effluent, with its large volumes of highly coloured water, has been a source of concern to the industry for years. It is the anionic reactive dyes and some acid dyes that cause most problems. A variety of means used to treat this effluent i.e. chemical, biological or physical are available (Robinson et al, 2001). These dye molecules however, pass through a conventional activated sludge system at the STW virtually unchanged (Wilmot et al, 1998) and now in the days of 'polluter must pay' principle, the onus is on the dyers to sort the problem out at source rather than leaving it for the STW to find the solution.

Chemisolv Ltd produces coagulants and flocculants for industry where there is a particular problem with colour or other contaminants in an effluent wastestream. Chemisolv CB300 is a blend of chemicals, with aluminium sulphate $(Al_2(SO_4)_3)$ as a major constituent plus a polyelectrolyte that is blended to suit a customer's own particular effluent treatment problem. Chemisolv CB300, the particular blend for Strines Textiles, was added as the treated effluent left the aeration pond prior to being transported by the Archimedean screw to the settling tank. It was added to aid flocculation of the colour laden flocs in the tank before the sludge flocs were transported to either the "Return Activated Sludge" (RAS) pipe or the aquabelt for dewatering.

EXPERIMENTAL

Activated sludge was collected from the "Return Activated Sludge" (RAS) pipe at Strines Textiles and taken into the laboratory. The "Mixed Liquor Suspended Solids" (MLSS) value was measured as 1.5% (w/v). This is much higher than the working MLSS at Strines which was approximately 0.5% (w/v) under normal working conditions. An activated sludge pilot plant was set up as described in earlier work (Smart et al, 2000) with a working solids content of 0.5% (w/v). Aliquots, 200cm^2, were taken from this pilot plant to treat dye solution and measure the uptake of colour by the activated sludge flocs.

The dyes used were Procion Reactive dyes supplied by BASF in their commercial form, Blue HEGN 125 and Crimson CXB. Dye solutions were made up as 1% (w/v) stock solutions in distilled water. Dye solutions were subjected to hydrolysis to mimic the spent reactive dyestuffs found in a textile effluent wastestream. The hydrolysis reaction was initiated by adding Na_2CO_3 (10g) to dye stock solution (500cm^3), the temperature of solution was increased to boiling point and maintained at that temperature for thirty minutes. Each stock solution was allowed to cool, placed in a stoppered volumetric flask, wrapped in aluminium foil and kept in a darkened cupboard to avoid any degradation of the dye by the light and to minimize evaporation.

Mini pilot plants were set up with 0.5% suspended solids, (using either Strines or Ringley Fold sludge), pH of 6.5 ± 0.5 and aerated to give an oxygen concentration of 7 mgdm$^{-3} \pm 0.5$. Dye stock solution 1%, volume 0.7 cm^3 was added to each plant to give a dye concentration of 3.5×10^{-3} % (w/v). To each plant varying quantities of Chemisolv CB300 (supplied by Strines Textiles) were added to give the following concentrations (with 0% to be used as the control): 0.1%, 0.25%, 0.5%, 1%, 2.5%, 5% and 10% (w/v). Absorbance readings to monitor the colour removal were taken at 0, 15, 60 and 180 minutes and from this % colour removal calculated, as shown by the following equation:

$$\% \text{ Colour removal} = \frac{At_0 - At_n}{At_0} \times 100$$

where,

At_0 = absorbance at time = 0

At_n = absorbance at time = t_n

All measurements were carried out on a double beam Camspec M350 UV/VIS spectrophotometer. Aliquots, (10cm^3) were taken from the mini-plants at intervals during treatment and centrifuged for five minutes at 5000rpm using a Camlab 1020 centrifuge supplied by Centurion Scientific to remove any turbidity which might have otherwise caused absorbance reading errors. All absorbance readings were taken at the original wavelength of maximum absorbance, λmax, for each dye against a sludge supernatant blank.

RESULTS AND DISCUSSION

Chemisolv and Strines sludge

The degree of colour removal using Strines sludge and variable concentrations of Chemisolv can be seen in Table 1 with the graphs of colour removal versus time in the presence of Chemisolv shown in Figures 1 and 2. Colour removal in excess of 80% is recorded in the mini-plant with 0% Chemisolv, as would be expected with Strines sludge which has been conditioned over the years to treat textile waste from dyeing and printing operations and the activated sludge plant has adapted over the years to treat a narrow range of pollutants very successfully. Chemisolv has been used in unusually high concentrations in these experiments to assess its effects on colour removal.

Table 1. % Colour removal at variable Chemisolv concentrations, Strines sludge

Dye	Time (mins)	Chemisolv concentration %						
		0	0.1	0.25	0.5	1	2.5	10
Blue	0	0	0	0	0	0	0	0
HEGN	15	86	96	109	109	102.6	35.4	-26.7
	60	87	95	104	108	102.9	32.3	-54.1
	180	79	92	106	106	100.3	31.9	-59.6
Crimson	0	0	0	0	0	0	0	0
CXB	15	78	102	109	107	84	28.4	-2.3
	60	83	104	108	108	85	32	-10.1
	180	83.5	100.5	106	105	81	32	-24.4

In the presence of the highest concentration the negative figures of -24.5% for Crimson CXB and –59.6% for Blue HEGN reveal that there has been an increase of colour at λmax within the supernatant in the treatment plant. The implication from this is that the Chemisolv is responsible for releasing colour from the sludge floc that has already previously been adsorbed prior to the set-up of the treatment plant. It is not unusual for residual colour within the sludge floc to be released. During the set-up of a pilot treatment plant the supernatant would become coloured, usually purplish, within 24 hours. The pilot plant was always set-up using sludge flocs and distilled water; therefore any colour could only come from the flocs. Strines industrial activated sludge has already been used for treating textile effluent containing reactive dye waste, therefore the flocs would hold some already adsorbed colour within them.

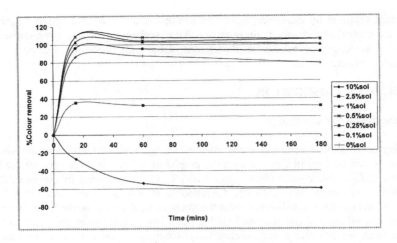

Figure 1 % Colour removal versus time Procion Blue HEGN using Strines sludge with variable concentrations of Chemisolv

When large concentrations of Chemisolv are used, i.e. 10% concentration, which is much larger than would ever be used in industry, instead of coagulation of colour and entrapment within the sludge flocs by either physical or chemical means, Chemisolv could cause the colour to be dispersed through the supernatant. Instead of allowing coagulation and flocculation to occur the large concentration is causing the sludge floc to break down and release colour already held within it. The colour removal in the presence of 2.5% Chemisolv is shown to be between 30 – 35% for both dyes. This suggests that at this concentration the Chemisolv is indeed holding the colour in the supernatant rather than allowing it to be adsorbed on to the sludge floc.

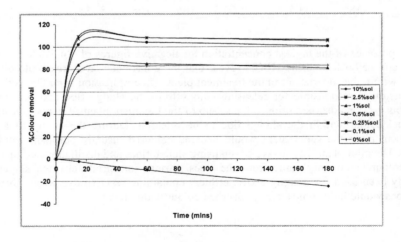

Figure 2 % Colour removal versus time Procion Crimson CXB using Strines sludge with variable concentrations of Chemisolv

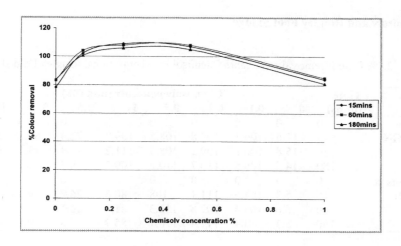

Figure 3 %Colour removal versus Chemisolv concentration with time, Blue HEGN, Strines sludge

Figures 3 and 4 show Percentage Colour removal versus Chemisolv concentration with time and it can be seen that again there is an optimum Chemisolv concentration. The optimum Chemisolv concentration is taken to be that at which 100% of the reactive dye colour is removed from the supernatant. Observation of the plot in Figure 3 shows the optimum Chemisolv concentration for Blue HEGN to be approximately 0.18% whereas Figure 4 reveals an optimum concentration of 0.2% for Crimson CXB.

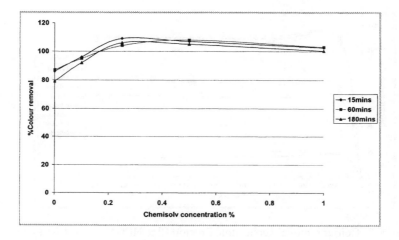

Figure 4 % Colour removal versus Chemisolv concentration with time, Crimson CXB, Strines sludge

Table 2. % Colour removal at variable Chemisolv concentrations, Ringley Fold sludge

Dye	Time (mins)	Chemisolv concentration (%)							
		0	0.1	0.25	0.5	1	2.5	5	10
Blue	0	0	0	0	0	0	0	0	0
HEGN	15	17.6	106.1	109.8	108.5	109.2	72.4	68.7	62.9
	60	15.8	106.1	109.2	108.5	111.2	73.4	73.1	62.6
	180	14.5	106.5	110.7	110.2	109.2	78.3	76	65
Crimson	0	0	0	0	0	0	0	0	0
CXB	15	5.3	107.6	111.1	108	80.5	26.2	9.8	-2.1
	60	6.6	105.4	107.8	112.3	81.9	26.1	3.8	-1.6
	180	5.3	104	111.4	116.1	85.7	25.8	9.5	-3.9

The colour removal results obtained when dye and Chemisolv are introduced into mini-pilot containing Ringley fold sludge are displayed in Table 2. The data are plotted graphically in Figures 5 and 6 showing % colour removal versus time with varying concentrations of Chemisolv CB300 for Procion Blue HEGN125 and Crimson CXB using Ringley Fold sludge. The colour removed by the sludge in the absence of Chemisolv (i.e. the control comparator) is in keeping with the results discussed in earlier work (Smart, 2004). The Ringley Fold sludge alone is only able to remove small quantities of the reactive dye colour from the treatment plants with the results indicating colour removal of 18% for Blue HEGN and 8% for Crimson CXB, this can be seen from Figures 5 and 6 respectively.

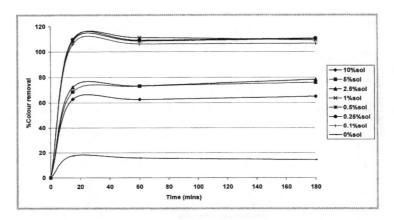

Figure 5 % Colour removal versus time Procion Blue HEGN using Ringley Fold sludge with variable concentrations of Chemisolv

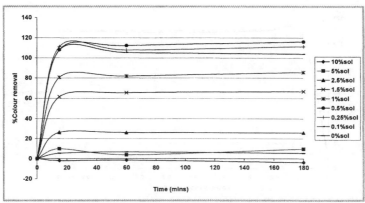

Figure 6 % Colour removal versus time Procion Crimson CXB using Ringley Fold sludge with variable concentrations of Chemisolv

Any concentration of Chemisolv higher than 0.5% reduces the quantity of colour removed from the treatment plant. It is apparent that there is an optimum Chemisolv concentration that gives maximum colour removal, beyond which its effectiveness at colour removal is reduced. A critical biosolids concentration was observed in the industrial sludge (Smart, 2004). Above this critical concentration colour removal was still effective but impaired. It was considered that this was due to an increase in floc size that reduced the overall surface area to mass ratio, thus decreasing the number of adsorption sites available on the surface for the dye. The effect could be similar here, at a critical Chemisolv concentration larger flocs are produced with reduced surface area to mass ratio and reduction in number of adsorption sites, leading subsequently to a reduction in the amount of colour removed.

In the Strines sludge it would appear that this phenomenon is occurring until, as stated earlier, there is a probable breakdown of the sludge floc with a release of the colour held within the floc. This does not seem to happen with the Ringley Fold sludge to the same extent, where there is a very slight increase of colour in supernatant in the mini-plant with Crimson CXB. The sludge flocs in the plant with Blue HEGN are still removing significant amounts of dye at 10% Chemisolv concentration although the colour removal is less than that at 0.1% Chemisolv concentration.

The optimum concentration can be seen clearly from Figures 7 and 8 which show % colour removal versus Chemisolv concentration with time for Blue HEGN and Crimson CXB respectively. The critical concentration appears to be at around 0.1%. The colour disappearance is rapid and results in coloured flocs. At the concentrations between 0.1 and 0.25% the colour disappears from the flocs during the three hour period of the treatment. The sludge flocs were examined optically after the supernatant was centrifuged off for absorbance measurements. The colour could be seen on the floc after 15 minutes reading but by 180 minutes the colour had disappeared. This was only noticed at the low concentration range. An explanation for the coloured flocs could be that initially the dye is attracted to the outside of the floc by chemical means or physical entrapment.

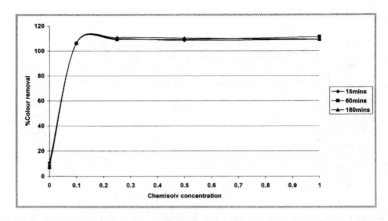

Figure 7 %Colour removal versus Chemisolv concentration with time, Procion Blue HEGN, Ringley Fold sludge

Over the three hour time period the colour is dispersed throughout the floc and therefore the colour does not appear on the outside of the floc.

The Ringley Fold sludge alone is only capable of removing less than 15% colour, while complete colour removal is apparent at 0.1% Chemisolv (although concentrations of Chemisolv up to 5% do remove a significant amount of the colour). Figures 7 and 8 indicate that the optimum Chemisolv concentration for 100% colour removal of both Procion Blue HEGN and Crimson CXB is slightly less than 0.1%, which would fit in with the data collected from extensive studies by Churchley (1999), where the optimum concentration was found to be 0.08%.

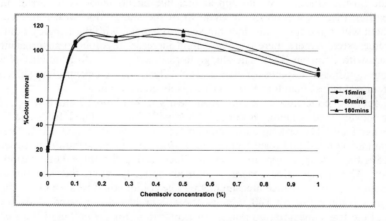

Figure 8 %Colour removal versus Chemisolv concentration with time, Procion Crimson CXB, Ringley Fold sludge

Chemisolv in the sludge floc

It was important to establish if the Chemisolv was present within the Strines sludge floc in a significant amount to be responsible for the success of the Strines sludge at removing colour. The λ_{max} was monitored in the presence of and without Chemisolv using both sludge supernatant and distilled water.

Table 3. λ_{max} shift for Procion Blue H-EGN and Crimson CXB in the presence of Chemisolv.

	λ_{max} (nm) Blue HEGN	Shift nm	λ_{max} (nm) Crimson CXB	Shift nm
Dye + water	626		543	
Dye + water + 1%Chemisolv	649	23	560	17
Dye + Strines supernatant	624		544	
Dye + Strines supernatant + 1%Chemisolv	646	22	560	16

The results shown in Table 3 indicate a bathochromic shift in the absorption spectra and consequently λ_{max} for both dyes, the shift being 22nm and 16nm for Blue HEGN and Crimson CXB respectively. The fact that no change in λ_{max} was observed for either dye in the pilot plants made with "Return Activate Sludge" from Strines is a good indication that any Chemisolv present is negligible or if present is possibly 'locked' within the sludge flocs and unable to participate in the colour removal process. Therefore adsorption of the dye on to the biomass is likely to be a function of this particular conditioned activated sludge and not the result of residual Chemisolv. It can quite easily be seen that the presence of Chemisolv brings about a very significant bathochromic shift in the λ_{max} for both Blue HEGN and Crimson CXB.

The data shown in Table 4 indicates a shift in the adsorption bands at very low concentrations of Chemisolv, as low as 0.01% (v/v). There is evidence of a significant bathochromic shift in λ_{max} at this low concentration. The shift for Blue HEGN is 34nm at 0.01% and above, for Crimson CXB the same bathochromic shift of 20nm is given at 0.01% and above. This shift in λ_{max} is significant and is a good measure of very low

Table 4. Shift in λ_{max} for Procion Blue H-EGN and Procion Crimson CXB in aqueous solution with varying Chemisolv concentration.

[Chemisolv] %	λ_{max} (nm) Blue HEGN	Shift nm	λ_{max} (nm) Crimsons CXB	Shift nm
0	626		544	
0.002	627	1	545	1
0.010	660	34	564	20
0.025	660	34	564	20
0.050	660	34	564	20

concentrations of Chemisolv present within the RAS. In the absence of Chemisolv each dye solution was added to distilled water to give the correct measurement concentration and the same procedure was repeated with Strines supernatant. When absorbance spectra were run on these two solutions, λ_{max} was recorded as the same for each dye in both solutions. Since it has been established that a concentration of Chemisolv as low as 0.01% caused a significant shift at λ_{max}, this would suggest that insignificant amounts of Chemisolv are found in the Strines sludge floc, from which the supernatants are prepared for colour measurement

CONCLUSIONS

The addition of Chemisolv into the mini-treatment plants improved the colour removal properties of both sludges although the largest increase was seen the Ringley Fold sludge. The Ringley Fold sludge alone removed less than 20% of the total colour for both dyes but on addition of the optimum Chemisolv dose of 0.1% (v/v), colour removal of 100% is recorded. Total colour removal increased in Strines sludge from approximately 80% to 100% with the addition of Chemisolv. A significant bathochromic shift in the λ_{max} of each dye in the presence of low concentrations of Chemisolv show that there are no significant residual levels of Chemisolv in the Strines RAS.

REFERENCES

1 Anon, 'The fibre Year' *Textile Month*, September 2002 36-39.

2 J Churchley, (1999), Private report to Chortex Ltd., Bolton.

3 C Diaper, V M Correia, S J Judd, 'The use of membranes for the recycling of water and chemicals from dyehouse effluents: an economic assessment' Journal of Society of Dyers and Colourists, 1996, 112, October, 273-281
Pierce J, 'Colour in textile effluents – the origins of the problem' *J of Society of Dyers and Colourists*, April 1994 **110** 131-129.

4 T Robinson, G McMullan, R Marchant, P Nigam, 'Remediation of dyes in textile effluent: a critical review on current treatment technologies with a proposed alternative' *Bioresource Technology*, 2001 **77** 247-255.

5 G Smart, 'Bioremoval of recalcitrant colour from effluent wastestreams' PhD Thesis, Bolton Institute, 2004.

6 G Smart, J Binkley, M Lomas, 'The effect of additives on effluent treatment of reactive dye colour removal' Wastewater treatment: Standards and technologies to meet the challenges of 21st century, Leeds, Terence Dalton, 2000.

7 N Wilmot, J Guthrie, N Gordon, 'The biotechnology approach to colour removal from textile effluent' *J of Society of Dyers and Colourists*, February 1998 **114** 38-41.

THE ANAEROBIC DIGESTION OF TEXTILE DESIZING WASTEWATER

Richard Dinsdale [(1)], Kevin Bryne[(2)] and David Tucker[(2)]

[(1)]Sustainable Environment Research Centre, University of Glamorgan,
Pontypridd, UK
[(2)]Water Development Services, 142 Whitchurch Road, Cardiff, UK.

ABSTRACT

Anaerobic digestion is a biological process which converts the chemical oxygen demand (COD) present in wastewater to methane and carbon dioxide and thus can be used to reduce the treatment costs of industrial wastewaters. Textile desizing wastewater is a high strength wastewater (20,000 mgl^{-1} COD) and extremely alkaline (>pH 12) and therefore is costly to discharge to sewer and will require neutralisation to a lower pH before discharge to sewer. An anaerobic treatment stage consisting of two 50 m^3 mesophilic upflow anaerobic sludge blanket reactors (UASB) with an activated sludge polishing stage was built and operated from August 2003. This wastewater treatment plant has contributed to a COD reduction of up to 80% and a neutralisation of the pH from pH 12 to pH 7-8. The methane present in the biogas is utilised to heat the anaerobic digesters and the CO_2 produced in the biogas is used to neutralise the alkali in the wastewater. This results in an in-house wastewater treatment process with low running costs i.e. low in energy demand and chemical usage.

INTRODUCTION

In the UK with 96% of the population connected to the sewer system, trade effluent treatment is usually performed by the regional water service companies. Not only may the cost of this service be substantial but the water company may not be prepared to take certain types of waste due to the infringement of consent limits. Significant rises in trade effluent charges may also be expected to meet increased environmental standards such as the EU Water Framework Directive and other future environmental protection legislation. When the regional water companies were privatised in 1989 a pricing regime of "Retail Price Index (RPI)+ K" factor was legislated for to compensate for the increase investment required to meet increased water discharge standards. This was in contrary to other utility privatisations where a price regime of RPI – X (where X is a cost reduction factor) was imposed. Indeed the water companies have a further option to increase prices via the "pass through option" (Kinnersley, 1994). Therefore, any company discharging to sewer faces the prospect of trade effluent charges rising faster than inflation on a yearly basis. This rise in charges along with environmental legislation has increased the incentive for in-house effluent treatment. By the intelligent selection of physical and biological treatment systems, an effective and reliable on-site wastewater treatment system could be installed. This would result in a waste with a reduced treatment cost and acceptable for discharge to the water company. In many cases, anaerobic digestion can provide a better, faster and more energy-efficient effluent treatment process than other processes with typical cost savings on effluent discharge of 65% (DTI, 2001). Increasing interest has been shown in anaerobic digestion as a biological treatment system as it has significant potential advantages over aerobic treatment.

THE ANAEROBIC DIGESTION PROCESS

Anaerobic digestion is a biological process which utilises bacteria that operate without the need for a constant supply of oxygen. These bacteria convert COD to methane and carbon dioxide thus reducing the trade effluent charge for the effluent treatment. The Biochemical Oxygen Demand (BOD) of the trade effluent, although using aerobic organisms in the test assay, can be an indication of the degree of COD reduction which can be achieved by the anaerobic digestion process. However as the process differs fundamentally from the aerobic process specific anaerobic biodegradability testing on the trade effluent should be performed to give a truer indication of the degree of anaerobic treatment.

The advantages of anaerobic digestion over aerobic wastewater treatment are lower electrical power usage, the production of energy in the form of methane gas, and lower microbial cell (sludge) production. A comparison of the aerobic vs anaerobic treatment of a tonne of COD (chemical oxygen demand) removed is shown in Table 1.

Table 1. Comparison of the anaerobic and aerobic treatment of a tonne of COD.

Factor	Anaerobic	Aerobic
Electrical Input	—	1100 kWh
Methane	1.16×10^7 MJ	—
Net cell production	20-150 kg	400-600 kg

(Data From Speece, 1983)

Anaerobic digestion does have some disadvantages. Anaerobic digestion is not an all in one solution, post-treatment will usually be necessary as BOD is not usually reduced sufficiently for discharge to open water. In particular, nitrogen and phosphorus removal is limited. Industrial plants usually operate at 20-35°C although can operate efficiently down to temperatures of 10-15°C (Mergaert et al, 1992). This problem is usually easily overcome as most wastewaters are released at these temperatures and methane from the digestion process can be used to heat the reactor. However it means that for dilute low strength wastes (<500 mgCOD l^{-1}) discharged at less than 10-15°C anaerobic digestion may not be the most cost effective option.

Despite these disadvantages, the cost advantage for the anaerobic process can be substantial for wastes with COD strength greater than 500 mgCOD l^{-1} and this cost advantage increases as the COD strength increases. For a waste of 2000 mgCOD l^{-1}, anaerobic treatment is a third of the cost of aerobic treatment and if the COD rises to 12,000 mgCOD l^{-1} anaerobic treatment of the waste can be a tenth of the cost aerobic treatment. With increasing energy and sludge disposal costs the advantages of anaerobic digestion can only improve (DTI, 2001).

DESIZING WASTEWATER CHARACTERISTICS

Sizing compounds such as starch and polyvinyl alcohol (PVA) are applied to yarns to impart tensile strength before weaving. These size compounds must be removed before the cloth can be dyed or printed. Textile desizing effluent is produced from the processes used to remove these size compounds prior to the cloth being dyed or printed. A typical desizing process utilises sodium hydroxide/hydrogen peroxide mixture to

164

hydrolyse the starch size and bleach the cloth with a subsequent washing stage at 80^0C to wash the hydrolysed size from the cloth. This desizing process results in the production of a significant volume of hot wastewater $(65-75^0C)$, with high COD of up to 20,000 mgl^{-1} and is extremely alkaline (>pH 12). In the S. Wales region this would incur a trade effluent charge of approximately £5.70 m^3 plus neutralisation costs.

THE DESIGN PROCESS

Laboratory studies

To aid in the selection of the most suitable on-site biological treatment process laboratory trials were performed at the University of Glamorgan using a 5 litre biological aerated filter (BAF) and 5 litre UASB anaerobic reactor. Using two batches of desizing wastewater the BAF was found to give at best a 50% removal in COD whereas the UASB reactor gave up to 90% COD removal with an average of 74% COD removal over a period of 25 days of continuous operation. The residual BOD from the anaerobic reactor was found to between 200-800 mgBOD l^{-1}. Ultimate aerobic biodegradability was found to be 80% of the total COD. The high strength of the effluent indicated that anaerobic digestion would be the most cost-effective solution. Although anaerobic digestion would be the main COD removal process utilised, an aerobic polishing stage should be implemented to further reduce the COD and remove the dissolved methane from the effluent discharged to sewer.

Neutralisation of the waste from pH 12 to pH 7 was also successfully achieved by the addition of carbon dioxide. Carbon dioxide neutralisation offers a number of advantages over mineral acid neutralisation. The main advantage is that as the anaerobic digestion process produces carbon dioxide, in-situ neutralisation using this carbon dioxide could be implemented without incurring the cost of the mineral acid dosing and consumption.

Full Scale Implementation

Once the decision to use anaerobic technology was taken, a survey of the factory's desizing process was conducted to determine the current production volume and strength of the desizing effluent and an indicative value for future production volumes. This data was then used to size the biological treatment process. Anaerobic digesters are available in a number of configurations e.g. CSTR (continuously stirred tank reactors), UASB reactor etc. A selection of UASB technology was made for a number of reasons, i) the waste characteristics met the design criteria for treatment in a UASB,ii) the laboratory studies were performed using a UASB reactor and iii) the UASB is a widely used technology. The waste survey data was used to size the digesters, using the design criteria described by Lettinga and Hulshoff Pol (1991). Therefore a maximum OLR (organic loading rate) of 10 kg m^3 day^{-1} with a HRT (hydraulic retention time) of 2 days was selected, giving a total reactor volume requirement of 100 m^3. For ease of build and a degree of redundancy within the system the reactor volume was divided in to two UASB reactor systems of 50 m^3 total volume. Using the information contained in Lettinga and Hulshoff Pol (1991), the two UASB reactors were designed and constructed in GRP (glass reinforced plastic). The only modifications made to the standard UASB reactor design was the implementation of an 8m^{3}h^{-1} recycle rate to improve in reactor mixing to aid in temperature control, to reduce the impact of the high strength effluent on the granular bed and to aid the carbon dioxide neutralisation process. To balance out the wastewater flow from the factory and to try to ensure a balanced flow to the UASB reactors over a weekly period, a 200m³ balancing tank was built into the system (see Figure 1). As the aim of the project was to ensure a process with a no or little energy or chemical inputs, and to ensure optimum usage of the biogas, a 50 m³ gas storage system was also implemented.

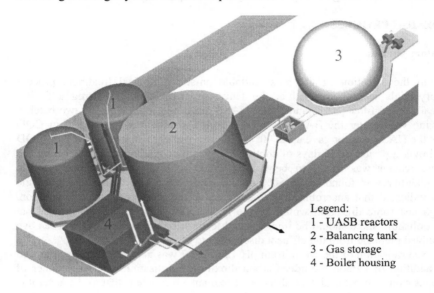

Legend:
1 - UASB reactors
2 - Balancing tank
3 - Gas storage
4 - Boiler housing

Figure 1   Layout of the Biological Treatment Plant at Treforest Textiles.

INITIAL RESULTS FROM FULL SCALE OPERATION

The digesters were seeded in August 2003 with anaerobic granules from a UASB reactor treating citric acid wastewater. The digesters are maintained at between 30-35^{0}C largely using the biogas produced by the anaerobic digesters. From the 8 months after digester start up an OLR of 8 kg m³ day^{-1} with a HRT of 2 days have been achieved. The effluent has been netralised to between pH 7-8 without the addition of any chemicals to achieve this. The COD removal percentage has varied between 40% and 80%.

FUTURE WORK

The future work is mainly directed at stabilising COD removals, this will be achieved by fine tuning the addition of nutrients, developing the best operational procedures for dealing with start-up after quarterly shut downs and dealing with peak loads and investigating potential inhibitors in the effluent such as hydrogen peroxide.

CONCLUSIONS

A low energy and chemical input anaerobic biological process has been developed and implemented. The effluent pH has been successfully neutralised to target levels with a reduction in COD being achieved. However with all biological treatment systems suitable operating and management procedures have to be developed and implemented over time to ensure optimal efficiency.

ACKNOWLEDGMENTS

The authors would like to thank the DTI and Treforest Textiles for their financial support to this project and Peter Jackson of Tate & Lyle for the supply of UASB reactor granules.

REFERENCES

1 DTI Industrial Wastewater and Effluent Treatment: A Review of AD Technology. *Bio-Wise Programme*, 2001.

2 D Kinnerlsey, 'Coming Clean: The politics of water and the environment', *Penguin Books*, London, 1994.

3 G Lettinga and L W Hulshoff Pol, 'UASB process design for various waste waters', *Wat Sci Tech,* 1991 **24**(8) 87-107.

4 K Mergaert, B Vanderhaegen, and W Verstraete, 'Applicability and trends of anaerobic pre-treatment of municipal wastewater', *Wat Res,* 1992 **26**(8) 1025-1033.

5 R E Speece, 'Anaerobic biotechnology for industrial wastewater treatment', *Environ Sci and Tech,* 1983 **17** 416-427.

EFFLUENT TREATMENT USING A SUBMERGED AERATED FILTER

Elaine Groom

A BIO-WISE Demonstrator Project, Applied Technology Unit, QUESTOR Centre (Queen's University Environmental Science and Technology Research Centre), Queen's University, David Keir Building, Stranmillis Road, Belfast BT9 5AG, Northern Ireland

ABSTRACT

In common with many companies in the textile sector William Clark & Sons Ltd produce a coloured wastewater that is highly variable in composition, which presents a challenge to treatment. The company recently chose to install a submerged aerated filter (SAF) to ensure a high rate of biological treatment in order to meet a river discharge consent. This technology, unproven in the sector, was awarded a BIO-WISE Demonstrator grant to optimise and demonstrate the effectiveness of the technology for textile effluent treatment and to assess the potential for its integration with other treatment technologies for colour removal and for treatment of wastes from a lamination process.

The plant was commissioned in 2002 and, following a period of detailed monitoring and optimisation, has consistently reduced the chemical oxygen demand (COD) and biochemical oxygen demand (BOD) of the effluent to levels required for discharge. The plant has also performed well in response to changes in effluent composition due to variations in day-to-day activities and the seasonal nature of the textile industry.

Project overview

The application of treatment technologies to textile effluent is a continuing problem. For companies needing to install or upgrade effluent treatment facilities, a number of choices exist of technologies proven to be applicable to textile effluent treatment. However, these tend not to include newer technologies that often represent a significant advance over existing processes. This is unfortunate as the textile sector contains a disproportionately high number of SMEs, a business category that can ill-afford to take risks with unproven technology, despite requiring access to the most cost-effective treatments.

Recent initiatives have tended to focus on colour removal due to changes in legislation that have allowed the introduction of colour consents for effluent discharges. For many companies however, levels of BOD and COD in effluent are a major concern as they contribute significantly to trade effluent charges. Where river discharge is possible, there is likely to be no one treatment that can deal with colour removal and remove BOD to consent levels. In comparison with other parts of Europe, the application of biological treatments to textile wastewater in the UK has been low. This has reduced the opportunities to advance the application of high rate biological treatments in this sector.

Submerged aerated filters are an example of advanced biological treatment processes that represent a significant advance over activated sludge plants and biotrickling filters but remain unproven for the treatment of textile effluents. Among the advantages of SAFs are the high organic loading rates achievable, the small footprint and reduced cost due to availability as packaged plant with reduced civil costs.

Biological treatments such as SAFs have little appreciable effect on many types of textile dye. However, the removal of BOD from the effluent may allow treatment with one of a number of technologies that would not be economic for colour removal in whole textile effluent. Thus it is possible that SAF technology will provide part of a complete treatment solution.

This project intended to demonstrate the treatment of textile effluent by a submerged aerated filter at William Clark & Sons Ltd. Funding provided an opportunity to co-ordinate the expertise of the technology provider (STG) and user (William Clark & Sons Ltd) and enable the technology to be optimised and proven under real conditions. An extensive monitoring programme was carried out jointly between the QUESTOR Centre and William Clark & Sons Ltd to produce a detailed record of effluent production and SAF performance in relation to company activities. A number of process variables and operating conditions were examined, within the range of operation of the plant, in order to optimise the performance. The company activities varied with season and produced extremes of effluent condition proving that the process is sustainable over the entire range of conditions that may be encountered. It was also demonstrated that the process can be rapidly brought into operation after shutdown periods, for example holidays, maintenance etc.

The objectives of the project were as follows:

- To design and install an advanced biological treatment plant based on SAF technology that can be optimised for the treatment of textile effluent.
- To demonstrate the robustness of the process to the variations in the effluent due to day-to-day variations in activity and the seasonal nature of the textile industry.
- To investigate the applicability of the technology to co-treatment of waste from an aqueous-based lamination process.
- To validate the sustainability of the process and examine its potential for integration with processes for colour removal.
- To demonstrate the performance of the technology to the textile industry through a dissemination programme.

Liaison between the QUESTOR Centre and STG also examined the potential for co-treatment of waste from an aqueous-based lamination process, a new and growing activity for William Clark & Sons Ltd, and the potential for integration of SAF technology with existing and emerging technologies for colour removal, including those supported by BIO-WISE. Dyeing activities at William Clark & Sons Ltd fall into two distinct seasons, therefore a range of appropriate tests at laboratory and pilot scale were carried out, running concurrently with other aspects. The project examined how a variety of treatments could be readily combined with biological treatment to give a complete, cost-effective solution to the treatment of textile wastewater.

The combined knowledge of the partners of biological treatments and the textile sector, and the specific experience of STG in the design and operation of SAF technology, provided an opportunity to optimise the configuration and operation of the technology for future application in the textile sector. The extended nature of the detailed monitoring programme ensured that data of sufficient quality was collected over a suitable timeframe to prove the technology for use within the textile industry.

EFFLUENT TREATMENT AT THOS. CHADWICK & SONS LTD

M. Madden[1] and M. Andrews[2]

[1]ENCO, Confederation of British Wool Textiles, Valley Drive, Ilkley, UK.
[2]Thos. Chadwick & Sons Ltd., Eastfield Mills, Dewsbury, West Yorks, UK.

ABSTRACT

Effluent with high COD, solids content and trace pesticide is generated by raw wool scouring. Although current methods of on-site effluent treatment reduce these contaminants before discharge the remaining concentrations are sufficient to cost the company a great deal of money for sewage treatment and to warrant severe limits on pesticide concentrations under Integrated Pollution Prevention Control legislation. It is highly likely that discharge costs will increase and pesticide limits will be lowered in the near future.

Trials with fast acting anaerobic digestion pilot plant have shown that the effluent can be further treated and that the additional use of effluent "polishing plant" can reduce contaminant concentrations to very low levels. Furthermore, the anaerobic system may provide useful methane gas with which to power the main processes on the site.

This paper shows the results of an industrially-based project that is likely to go on to become a full-scale solution to a difficult problem not only for this company but for the wool textile industry as a whole.

INTRODUCTION

The scouring of raw wool results in effluents that are rich in grease and solids and hence have a high Chemical Oxygen Demand. As well as these major components, there are traces of the process chemicals such as detergent and builders and traces of sheep dip pesticide residues. All these substances mean that wool scour effluent is subject to high discharge costs and to strict limits particularly with respect to pesticides.

Many techniques are used around the world to try to reduce the pollution potential of wool scouring effluent but in the UK, where discharge costs are extremely high and environmental legislation is strictly enforced, simpler techniques of treatment are not sufficient.

Currently the technique of choice is to coagulate then flocculate the effluent and then to separate the floc from the cleansed effluent by means of decanter centrifuge. This results in a cleaner effluent and a waste sludge that can both be discharged at a lower cost and within current environmental limits. However environmental limits are getting more and more stringent and economic pressures mean that mills are looking to further reduce costs where they can. Under these circumstances further treatment of the effluent is desirable.

Anaerobic digestion is a tried and tested method for treating high strength effluents worldwide. It is used in a few scouring mills around the world as the primary treatment for scour effluent but typically requires a lengthy residence time for the treatment to be successful. It has the advantages of producing only a minimal sludge since most of the organic component of the effluent is "consumed" by the biomass in the reactor and it produces methane as a by-product of that consumption which can be used as a source of power.

Past research efforts to introduce anaerobic digestion into the UK scouring industry have met with limited success. Digestion and methane production did occur but the reactors used quickly became blocked with the inorganic component of the effluent in the form of dirt and sand.

The purpose of this paper is to describe briefly trials that have been taking place at the Standard Wool (UK) Ltd wool scouring plant in Dewsbury, UK with the help of funding from the UK government's BIOWISE initiative. These trials have examined the use of anaerobic digestion of scour effluent and have utilised pilot plants designed to treat the effluent in a relatively short period compared to conventional anaerobic digestion. Pilot plant to "polish" the effluent after the main treatment has also been evaluated.

Trials have been funded for a period of two years and the project is now entering the stage where consideration of a full-scale system has begun and costing of the components is underway. Parts of the system remain confidential at this stage although once the project is completed dissemination will be much more detailed.

THE COSTS OF TREATMENT

The costs of discharge of industrial effluents to sewage treatment undertakers in the UK are governed by the use of the Mogden formula:

$$C = R + P + (Ot/Os)*B + (St/Ss)*S$$

where C = the discharge cost to sewer in pence/m3
    R = reception charge at the sewage treatment works in pence/m3
    P = primary treatment charge at the sewage works in pence/m3
    B = biological treatment charge at the sewage works in pence/m3
    S = sludge disposal charge at the sewage works in pence/m3
    Os = the biological strength of combined sewage in mg/l COD
    Ss = the sludge strength of combined sewage in mg/l Settled solids
All the above values are set by sewage undertakers for their area each year.
    Ot = is the measured strength of the discharge in mg/l COD
    St = is the measured sludge strength of the discharge in mg/l Settled solids

This formula enables easy calculation of how much an effluent will cost to discharge.

A wool scour processing 18,000T of raw wool per year at 2.5 litres of scour water per kilogram of wool would typically have effluents with Ot = 80-100,000 mg/l COD and St = 30-35,000 mg/l Set. Solids. Using the formula with the values set by Yorkshire Water for 2003-4 would give a discharge cost of between £1,428,955 and £1,735,939.

Currently used effluent treatment would be expected to reduce COD values by around 70% and solids values by 90%. Thus discharge at present might typically cost between £335,465 and £409,805 per year.

Clearly there is room to reduce these costs further and the trials aimed to show that a system incorporating anaerobic digestion of effluent could achieve a reduction of approximately 70% in costs.

PESTICIDES IN SCOUR EFFLUENT

The use of sheep-dip chemicals during the wool-growing season is extremely widespread. Sheep are attacked by a number of insect pests in all climates and dipping, spraying or back lining of sheep at some point is very often a necessity. Whilst most wool growing and exporting countries in the world take care to advise farmers to use the minimum amounts of chemical and at times that allow high levels of degradation of the chemical in the fleece to occur before shearing, most wools still contain measurable amounts of these chemicals. There are several types of pesticide in use for sheep dipping, but in the UK three types have attracted environmental limits because of their proven effects on aquatic environments. These are the organochlorines, the organophosphates and the synthetic pyrethroids. The first group is largely banned from use on sheep but such is their persistence in the environment that measurable traces can still be detected in wool from countries where the bans are in force. There still remains some wool exporting countries where use of this group is commonplace. The second and third groups are legally used all around the world. All these chemicals can enter the aquatic environment via the effluent from raw wool scouring and being lipophilic in nature are associated with the wool grease that is deliberately removed from the fibre during the scouring process.

In the UK, most on-site effluent treatment was originally installed to help reduce overall pollution load and thus reduce disposal costs but has had the secondary effect of reducing the micro-pollutant (trace pesticide) load by removing much of the grease from the effluent.

As legislative limits on pesticides in receiving waters have become stricter to the point where Environmental Quality Standards (EQS) are approaching analytical instrument detection limits, it has become much more important for raw wool scourers to ensure that the discharge limits imposed by the Environment Agency on their effluent discharges are adhered to. Any breaches must be reported to the Agency who ultimately has the power to order factory closure until the operator can assure the Agency that limits will be met in future.

Current means of effluent treatment at UK scourers tend to reduce pesticide loads by 70 – 80% depending upon the sophistication of the treatment. The introduction of an anaerobic process to the treatment could result in even greater reductions in pesticide discharges and one of the aims of the trial was to show whether improved pesticide removal could be achieved by this technology.

TRIAL EQUIPMENT

Two separate systems were chosen for the trial. Parques Ltd (Netherlands) via the project partner Aquabio Ltd (UK) and the second by project partner Puriflow Ltd (UK) supplied these reactors. Both were based on the upward flow system where the effluent is introduced to the bottom of a columnar reactor with the cleansed effluent eventually being extracted from the top. The Parques reactor had a granular blanket supporting the biomass through which the effluent flowed whilst the Puriflow had a plastic medium on which the biomass grew and the effluent flowed in between. Perhaps the most important difference between the reactors was the hydraulic residence time of the effluent in the reactor. The Parques reactor, originally designed for effluent somewhat weaker than that from raw wool scouring, had a designed residence time of 15 hours whilst the Puriflow reactor was based on a residence time of 3 - 4 days. Increasing or decreasing the re-circulation of the effluent within the system could of course alter both these times.

The reactors were fitted with controlled heating to maintain a temperature suitable for the growth of the biomass and with pumps to feed and re-circulate effluent and, if required add extra nutrients. The plants were also fitted with a methane gas collection and measurement system.

The diagram below shows the general treatment scheme for the pilot plants.

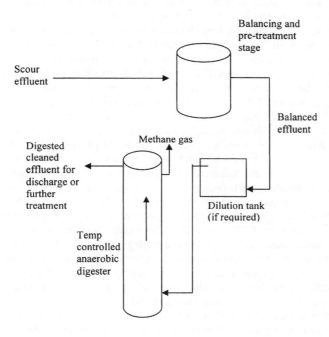

TREATMENT

Effluent from the scour was pumped to the balancing and pre-treatment section of the system and from here to a dilution tank.

Initially both reactors were fed with very diluted effluent so that the relatively high COD load would not overwhelm the biomass within them. As time went on it was possible to increase the concentration of the feedstock until the Puriflow reactor was accepting undiluted effluent and the Parques reactor was accepting 33% diluted effluent. After anaerobic digestion in the reactors, the effluent was discharged to drain. At a later stage in the project a pilot scale membrane bioreactor plant was installed to accept effluent from the Puriflow reactor and give further treatment. The membrane bioreactor worked with a combination of an aerobic biomass and a cross flow micro-filtration membrane to treat the relatively low solids and COD load effluent. It produced a concentrate that could be recycled back into the anaerobic feedstock and permeate which was discharged to sewer.

PERFORMANCE MEASUREMENTS

The intention from the beginning of the project was to try and run the reactors in such a way as to mimic the simplest form of full-scale operation. Whilst biological treatment plants can be made to work on almost any effluent, it is often not cost-effective to consider them due to the high on-going maintenance and technical manpower required to keep them in efficient working order. This project was set up to run the pilot plants on effluent that was as little modified as possible from that currently produced and to minimise interventions such as nutrient supplementation as much as possible. The only variables the team wanted to alter were feed concentration and feed rate; the object being to achieve the fastest rate at the highest concentration.

The performance criteria set by the team were equally simple. A successful trial would be one in which the pilot plant(s) resulted in at least a 70% reduction in overall pollution load across the digester, a significant decrease in overall load compared to the current technique and demonstrated a significant reduction in pesticides concentration in the final discharge. A second criterion was to reduce the amount of chemicals used in the overall treatment of the effluent. Current methods of treatment can use very large quantities of chemicals and the cost of treatment is driven upward by this. The production of methane gas was regarded as an extra benefit only if it was of sufficient volume and quality to be of use as a fuel for the scour process. The reactor(s) had to demonstrate "robustness" i.e. the ability to survive the various pollution loads delivered to it and to recover from any malfunctions that might occur on a full-scale plant.

COD and suspended solids content, as the main pollution load characteristics, were measured at very frequent intervals through out the trial duration. The wool grease content of the effluent was also measured frequently. The wool grease is one of the largest contributors to the COD load and is also a crude way of measuring approximate pesticide removal since the pesticides are largely dissolved in the grease. The concentrations of sulphate, phosphate and nitrogen/ammonia were measured on a regular basis so that the input and output of basic nutrients to the reactors could be followed. Pesticide analysis was also undertaken once it was considered that the reactors had stabilised at an optimum treatment rate.

The performance of the MBR had no set criteria, as it was uncertain as to the level of additional treatment it could deliver. The analysis of polished effluent was expected to show a significant decrease in both COD and residual pesticide content however.

RESULTS FROM THE PROJECT TO DATE

The results presented cover the following areas:
1) COD reduction over time for the two pilot plants - reactors only (Figures 1 and 2)
2) Grease reduction over time for the two pilot plants - reactors only (Figures 3 and 4)
3) Solids reduction over time for the two pilot plants – reactors only (Figures 5 and 6)
4) General COD reduction over the entire system – Puriflow reactor only (Table 1)
5) General Grease reduction over the entire system – Puriflow reactor only (Table 1)
6) General Solids reduction over the entire system – Puriflow reactor only (Table 1)
7) Pesticide reduction over the entire system – Puriflow reactor only (Table1)
8) COD, Grease, Solids and Pesticides reduction in comparison to current system (Table 2)
9) Gas production and quality
10) Performance of the MBR (Table 3)

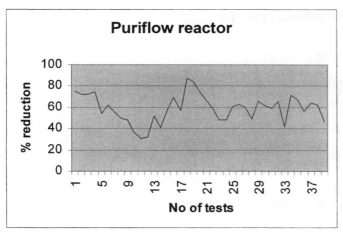

Figure 1 Percentage reduction of influent COD to Puriflow reactor

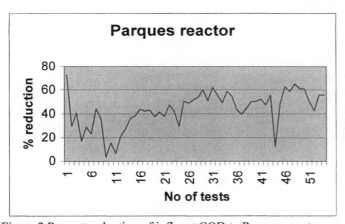

Figure 2 Percent reduction of influent COD to Parques reactor

As COD input to both reactors was increased over the trial duration the performance (reduction) of the Parques reactor appeared to improve but the Puriflow reactor gave the better performance considering the lower or zero dilution of its input.

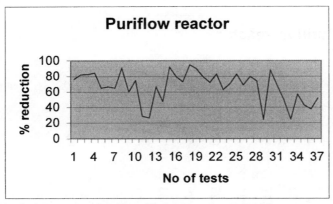

Figure 3 Percent reduction of influent grease to Puriflow reactor

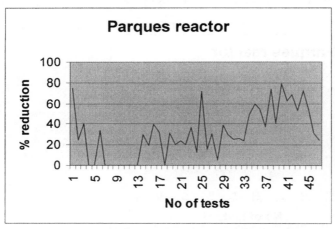

Figure 4 Percent reduction of influent grease to Parques reactor

The general trends show that grease reduction was more efficient in the Puriflow reactor despite the much higher concentration of the input. Points where the grease reduction for the Puriflow reactor dip below 40% are points at which practical problems with the pilot plant were experienced such pump breakdown or over-heating. The same is true for the point on the Parques graph where reduction seems to fall to zero.

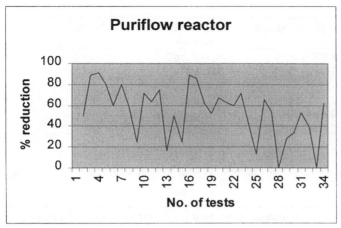

Figure 5 Percent reduction of influent solids to Puriflow reactor

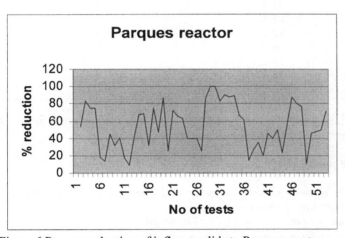

Figure 6 Percent reduction of influent solids to Parques reactor

The solids reduction in both reactors is much more variable than grease or COD reduction. The Parques reactor solids discharge actually mirrored the input closely whilst the Puriflow reactor performance has become less impressive as the input concentration has increased to 100%. These graphs do not reflect the problems experienced with solids breakthrough in the Parques reactor discussed more fully below.

Table 1. General reduction in COD, Solids, Grease and Pesticides across the whole system (from balancing tank input to Puriflow reactor discharge whilst operating at 100% effluent input.)

Parameter	Input to System	Discharge from Puriflow	% Reduction
COD	83,986 mg/l	8,749 mg/l	89.6
Set. Solids	30,679 mg/l	386 mg/l	98.7
Grease	1.59%	0.03%	98.1
OC Pesticide	9.053 ug/l	1.252 ug/l	86.2
OP Pesticide	697.43 ug/l	9.92 ug/l	98.6
SP Pesticide	252.65 ug/l	13.42 ug/l	94.7

Table 2. Comparison of reductions across the trial system with the current effluent treatment performance.

Parameter	Current Treatment Performance (% Reduction)	Trial System Performance (% Reduction)
COD	72.4	89.6
Set. Solids	85.1	98.7
Grease	88.3	98.1
OC Pesticide	68.9	86.2
OP Pesticide	76.2	98.6
SP Pesticide	25.6	94.7

(Current system values for the above parameters (input and output) are based on historical measurements for 10 typical working days)

Gas production and purity

Note that these results relate to periods during the trial when steady state treatment of effluent under minimum dilution was occurring.
Overall average volume of gas $= 0.3$ m^3/kg COD removed
Average purity of gas – CO_2 content $= 15\%$ v/v
$\qquad\qquad\qquad\quad$ H_2S content $= <1$ppm
$\qquad\qquad\qquad\quad$ CO content $= <0.05\%$ v/v
$\qquad\qquad\qquad\quad$ CH_4 content $= 80\%$ v/v
Balance of volume is made up by nitrogen.

Performance of the MBR system

Table 3. Reductions of pollutants as a result of using an MBR "polishing" system fed with effluent discharged from the Puriflow reactor.

Parameter	Input to System (ex Puriflow)	Discharge from MBR	% Reduction
COD	8,749 mg/l	3,065 mg/l	64.9
Set. Solids	386 mg/l	50 mg/l	87.0
Grease	0.03%	0.0094%	68.7
OC Pesticide	1.252 ug/l	0.205 ug/l	83.6
OP Pesticide	9.92 ug/l	0.45 ug/l	95.4
SP Pesticide	13.42 ug/l	0.49 ug/l	96.3

CONCLUSIONS FROM THE TRIALS

During the trials it has been clear that the Parques pilot plant with its short hydraulic retention time was not able to deliver the same reduction in overall load as the Puriflow plant. The graphs showing COD, Solids and Grease reduction with time show that the fast throughput system achieved reasonable results especially given the speed of treatment. However the plant only ever operated on diluted effluent feed and seemed unable to cope when dilution was reduced. For the purposes of the project, the aim was to select a technique of anaerobic digestion that could reduce its input load by 70% and thus improve the overall load reduction of the system in comparison to the current treatment. The Puriflow pilot plant came close to this target and is thought likely to achieve and even exceed it at full scale. The plant was linked with an MBR pilot scale system to measure whether any further worthwhile reductions in load could be attained.

Robustness was also an important selection criterion and during the trials it was clear that the Parques plant was more susceptible to changes in input load than was the Puriflow plant. Despite the use of a balancing tank to try to "smooth" out any sudden changes in effluent characteristics the short hydraulic retention time of the Parques plant meant that it had little time for its biomass to adjust to even slightly changing conditions. It was also shown that the Parques plant was more difficult to return to steady state running after practical problems such as temperature excursions in the biomass. One of the major problems with the Parques plant was, however, the distribution of solids within the reactor. The design of the effluent circulation within the reactor itself meant that solids were lifted from the lower parts of the reactor column by rising bubbles of gas and became lodged in the gas take-off ports. This meant that the reactor had to be regularly cleaned and it became clear that the short residence time and the reactor circulation could not cope with the effluent strength despite prior dilution.

The longer the period of time that can be allowed for digestion, the more effective an anaerobic reactor system is likely to prove but it is obvious from the results that some portions of the feed effluent are more easily digested than others. Grease reduction across the whole system (Puriflow reactor) is excellent and is significant for both reactors measured in isolation but there is clearly another component of the effluent COD that is not being digested so quickly. This is highly likely to be suint. Suint is essentially the sweat of the sheep dried on to the wool that is to be scoured. It is made up principally of organic potassium salts that have a characteristic dark brown colour in solution. This coloration is evident in the discharge from the anaerobic reactors and is indicative of suint presence. Being water-soluble it likely that the salts would need to be exposed to the anaerobic conditions for longer than used in these trials in order to be digested or degraded.

The solids in the effluent are to some extent removed by the anaerobic digesters but the most effective part of the system in this respect is the balancing and preparation technology. Without solids reduction in front of the reactors it would not be possible to run the anaerobic components at such a relatively high rate of through-put.

The pesticide removal of the pilot system is excellent and is a significant improvement over the current treatment plant. It is known that the organisms within a biomass will tend to adapt to digest and degrade the substances they are regularly exposed to and again a longer retention time will aid in this process of adaptation. In the anaerobic reactors the biomass is breaking down the grease by digestion and the pesticides are suffering the same fate.

There has been no evidence of "breakthrough" of grease, pesticides or solids from the Puriflow reactor during its running period and this indicates that there is no or only very gradual build-up of sludge (undigested material) in the reactor. The structure of the reactor will allow, on a full-scale plant, the removal of small quantities of sludge build up in the base by simply opening a valve to remove consolidated material in the base. This material may then be de-watered.

The gas produced by the Puriflow reactor was remarkably high quality and consultation with CHP engineers confirmed that the gas could be used to power a CHP engine that would supply electricity to the main process or supply green electricity to the grid. Green electricity attracts favourable tariffs and this could provide some income to offset running costs of the effluent plant.

The MBR system did produce some excellent results especially in the area of pesticide concentration reduction. However the COD reduction was relatively poor and it was noted that the suint content of the effluent from the MBR was similar to that of the influent feed. It could be that membranes with reduced pore size might help solve this but the throughput rate of the MBR would decrease in this case.

THE POTENTIAL FUTURE

If the full-scale version of the pilot plant was to perform in the same way as described in this report then the following cost comparisons could be made:
(note: current system average discharge COD = 21,020 mg/L and SS = 2,042 mg/L)

Current treatment – discharge to sewer cost = £6.35/m^3
New system – discharge to sewer cost = £2.72/m^3
New system with additional MBR - discharge to sewer cost = £1.21/m^3
For a discharge of 45,000m^3 per annum (18,000T wool at 2.5L/kg discharge rate) the above values translate into:
Current treatment = £285,750 per annum (+ £65,000 current chemical costs)
New system = £122,400 per annum - an annual saving of £163,350 (not inc. chemicals)
New system + MBR = £54,450 per annum- an annual saving of £231,300 (not inc. chemicals)
The value of gas produced can only be crudely estimated by scale-up but would be in the region of £ 30 – 40,000 for a 45,000m^3 discharge.

There is thus considerable scope to cover the cost of some of the installation of such a new system despite the fact that such a system would be relatively expensive because of its uniquely custom designed and built nature.

Once costs have been covered (estimated pay back about 4 – 5 years for the option without MBR) the new system would immediately begin to add its cost savings to the company bottom line and would in addition improve the environmental profile of the company by reducing pesticide discharges and by producing green electricity or reducing on-site imports of electricity.

PROTECTIVE PROPERTIES OF TEXTILES DYED WITH NATURAL DYES

Deepti Gupta

Department of Textile Technology, Indian Institute of Technology Delhi, Hauz Khas, New Delhi-110016, India

INTRODUCTION

Now and in the future consumers will accept only special textile materials; these are textiles which enable differentiation of products, are multi-functional, technologically perfect and are available at a reasonable price. Today's customers are becoming more and more aware of problems caused by microbes - from the meat we eat to the clothes we wear. Exposure to sun is leading to increased incidence of skin cancer. Therefore claims for protection are increasingly. Researchers are now looking for natural products, which do not have any ill-effects on the general population and are easily degradable. Consequently, there is increased research activity to find solutions based on natural products for most protective finishes.

Interestingly, most of the plant materials used for natural dyes are also attributed to have medicinal properties. Hence, it was decided to undertake systematic research to study the special properties of natural dyes after they have been applied on textiles. In preliminary studies carried out by the authors [1] these dyes had shown good absorption in the ultra violet region of electromagnetic radiation. Hence, they could be expected to provide good sun protection (UPF).

There is no information available on the protective properties and behaviour of these products once they are dyed on to textile substrates. In this study, a series of experiments was conducted to screen the commercially natural dyes for UV absorbing and anti microbial activity and to study the effect of various dyeing parameters on these properties.

MEDICINAL PROPERTIES OF DYE PLANTS

Many natural dyes form a part of the Indian pharmacopoeia and are known to be effective antioxidants, anti-tumor and anti-diabetic agents. This paper presents a brief review of the reported medicinal properties of some plants which are known to be important sources of dye.

Ahmad and Beg [2] tested the ethanol extracts of 45 Indian medicinal plants traditionally used in medicine for their anti microbial activity against certain drug-resistant bacteria and a yeast *Candida albicans* of clinical origin. The plant extracts were tested against *Staphylococcus aureus, Salmonella paratyphi, Shigella dysenteriae, Escherichia coli, Bacillus subtilis* and *Candida albicans*. Broad-spectrum anti-microbial activity was observed in 12 plants, *T. chebula* being one of them. Qualitative photochemical tests, thin layer chromatography (TLC) of certain active extracts demonstrated the presence of common photo-compounds in the plant extracts including phenols, tannins and flavonoids as major active constituents.

Studies have been carried out with plants belonging to *Acacia* family. Naik et al report that the extract of *A. catechu* contains catechu tannic acid, catechin and quercetin [3]. These compounds are responsible for the colour characteristics as well as imparting powerful astringent and anti-oxidant properties to the extract. Kambizi et al studied the methanol extracts of *A nilotica* and found that they show significant inhibition against Gram-positive and Gram-negative bacteria [4].

Alkaloid constituents obtained from the roots of **Berberis vulgaris** have been used as medicine in rheumatic and other chronic inflammatory disorders [5]. The same plant yields a fluorescent yellow basic dye which is very bright but fleeting. Another yellow/red dye of historical significance is **Carthamus tinctorius** or Safflower. The aqueous extract has antioxidant properties according to Hiramatsu et.al [6]. A potent antioxidant has also been isolated from the safflower oil cake making it a good candidate for development of anti-inflammatory drugs. [7].

Many home remedies based on **Curcuma longa** or turmeric are still used in India . Turmeric owes its characteristic yellow color to three major pigments; curcumin (50–60%), demethoxy curcumin (20–30%) and bis (demethoxy) curcumin (7–20%) [8]. Scientific evidence has shown that it possesses bactericidal, anthelmintic activity and reduces cholesterol levels when given orally. Nowadays, curcumin is used in clinical trials for AIDS patients [9,10]. The chemical investigation on volatile oils of *Curcuma longa* [11], indicated it to be a good antifungal material. Turmeric also has pronounced anti-oxidant and anti-inflammatory effects and is anticipated to act as an anti-tumur promoter [12].

Rheum emodi is an important medicinal plant, which finds an extensive use in Ayurvedic and Unani systems of medicine. A recent publication [13], deals with a variety of biological properties of the compounds isolated from various *Rheum* species. A US patent discloses the insect-controlling activity of the ooze of *Rheum* genus, against sanitarily injurious insects [14].

Roots of **Rubia cordifolia** roots are a source of deep red shades on cotton. Coloring matter comprises of a mixture of hydroxy anthraquinones. Prominent among them are purpurin (trihydroxy anthraquinone) and munjistin. Roots are credited with tonic, astringent, antidysentric and antiseptic properties. They are used in rheumatism and form an ingredient of several ayurvedic preparations [15]. It has recently been credited with anti-tumur activity [16].

Quercus infectoria galls contain tannic acid (gallo tannic acid) as the principal constituent (50-70%) which are often used in dysentery and diarrhoea. [17]. A study by Redwane [18] et al. indicates the efficacy of extracts and fractions of *Quercus* galls as larvicidal agents and their possible use in biological control of *Culex pipiens*, the urban nuisance mosquito. **Terminalia chebula** contains tannin, gallic acid and chebulenic acid thus it acts as an effective purgative, astringent and blood purifier. The aqueous extract acts as an anti-oxidant, and protects the human body from liver cirrhosis, atherosclerosis, cancer etc [3]. It was found to inhibit lipid peroxide formation and to scavenge hydroxyl and super oxide radicals in vitro [19]. Bacterial species like *H. pylori E. coli*, *V.cholerae*, *V. vulnificus* etc were also inhibited by *Terminalia chebula* extracts.

Punica granatum (Pomegranate) is an important source of yellow dye. Researchers have reported the antibacterial, anti-fungal, anthelmintic and anti fertility activities of the various extracts of different parts of this plant [20,21]. Machado et al. [22] studied ethyl acetate extract of P*unica granatum* against multi resistant bacteria, *Staphylococcus aureus* by disc diffusion method. A mixture of ellagitannins isolated from the plant demonstrated antibacterial activity against all *S. aureus* strains tested. Significant activity was shown by the methanolic extract against *P. vulgaris* and *B. subtilis* [23].

MATERIALS AND METHODS

Scoured cotton fabric in plain weave was used for the study. Fabric thickness was 0.33 mm with a cover factor of 81.

Ten natural dyes procured from M/s Alps Industries Ltd, Ghaziabad, India were used for the study. The dyes used were *Rumex maritimus, Quercus infectoria, Mallotus phillipinensis, Rubia cordifolia, Rheum emodi, Kerria lacca, Acacia catechu, Terminalia chebula, Punica granatum* and *Acacia nilotica*. All chemicals used were of LR grade.

Dyeing and mordanting of test samples

Cotton fabric was dyed using 10% (owf) shade, at 80°C for 30 min using 1:30 MLR at neutral pH. Samples were also prepared with specific dye-mordant combinations as reported in Table I. Pre-mordanting, simultaneous mordanting - dyeing and post-mordanting were used. After dyeing, the samples were soaped with 0.5 gpl Lissapol N and rinsed in water.

Table I. Recipes used for dyeing of test samples

| Dye | Mordanting method | | | Time & temp |
	Pre-	Post-	Simultaneous	
A catechu	---	---	$CuSO_4$ 0.25%	80°C, 30 min
A catechu	---	---	$FeSO_4$ 2% $CuSO_4$ 2%	80°C, 30 min
R cordifolia	Alum 8%	---	---	80°C, 30 min
Q infectoria	---	$FeSO_4$ 0.5%	---	RT, 20 min
M philippinensis	Alum 8%	---	---	80°C, 30 min
Kerria lacca	Alum 10% Tartaric acid 5%	---	---	80°C, 30 min
Kerria lacca	---	---	$SnCl_2$ 2% Oxalic acid 10%	80°C, 30 min
Kerria lacca	---	$CuSO_4$ 1%	---	80°C, 30 min
R emodi	Alum 8%	---	---	80°C, 30 min
T chebula	Alum 8%	---	---	80°C, 30 min
P granatum	Alum 8%	---	---	80°C, 30 min
A nilotica	---	---	Alum 5%	80°C, 30 min
R maritimus	---	---	Alkaline dyebath $CuSO_4$ 2% TSP 5%	80°C, 15 min 80°C, 15 min 80°C, 15 min

Studies on UV-absorption of dyes

The UV spectra of dyes in solution were recorded using a Pharmacia Biochrom 4060 UV-Visible spectrophotometer. Scans were analysed for their absorption characteristics with a view to see if the absorption characteristics of dyes in solution could be used to predict the UV protection provided by dyes on fabric. On fabric, the transmittance measurements from 280-400 nm were taken on each specimen. Fabrics are assigned a UPF rating number and also a protection category depending on how much UV radiation they block out, as shown below in Table II:

Table II. UPF ratings and protection categories

UPF Rating	Protection Category	% UVR Blocked
15 - 24	Good	93.3 - 95.9
25 - 39	Very Good	96.0 - 97.4
40 and over	Excellent	97.5 or more

UPF values of fabric samples were measured on SDL UV protection measurement system based on a Campsec M350 UV-Visible spectrophotometer, using the standard AS/NZS 4399:1996. This system is equipped with an integrating sphere for diffused transmittance measurement and calculates the UPF value directly.

Determination of anti-microbial activity of selected natural dyes

The bacterial strains for testing were procured from MTCC (Microbial Type Culture Collection) Chandigarh. Testing for antimicrobial activity of dyes was carried out using AATCC test method 100 (colony counting method). The commercial antimicrobial agent Fabshield AEM 5700 was obtained from Rossari Biotech India Pvt Ltd, to be used as a reference.

RESULTS AND DISCUSSION

UV absorbance of dyes in solution

In evaluating sun protection efficiency, the critical zone is that between 280-320nm or the UV B region. These are high intensity rays which are known to cause skin carcinomas following extended exposure. UVC rays (200-280 nm) are absorbed by the ozone layer and do not reach the earth, while UVA (320-400) does not penetrate deep into the skin and is less harmful. To study the efficiency of dyes, their absorbance spectra in UV-Vis region was recorded. Shapes of the curves and areas under the peaks in various UV regions were analysed. Results are reported in Table III.

Table III. Analysis of the UV specra of natural dyes in solution

Dye	λ max (nm)	% of total area under the curve		
		UVA	UVB	UVC
R. maritimus	280	31.4	28.3	40.2
Q. infectoria	270	19.9	30.1	50.0
M. philippinensis	280	37.1	27.3	34.3
R. cordifolia	260	35.0	26.8	38.1
R. emodi	270	35.2	26.4	38.5
K. lacca	295	35.6	26.5	37.8
A. catechu	280	32.5	23.1	44.4
T. chebula	275	27.1	29.3	43.6
P. granatum	275	31.0	21.7	47.2
A. nilotica	280	49.2	22.4	28.4

It can be seen that λ_{max} for five dyes namely *R . maritimus, M. philippinensis, K. lacca, A. catechu* and *A. nilotica* lies in the UV-B region. All dyes absorb approximately 20-30% radiations in UV-B region, which is adequate for providing good UV protection. Hence it can be expected from this analysis that the selected natural dyes can act as good UV absorbers and are suitable for use in sun protective clothing.

UPF values of cotton dyed with Natural Dyes

Absorption characteristics of dyes in solution can be but indicative of behaviour or fabric. To study the actual protection provided, cotton fabric was dyed with two natural dyes *Q. infectoria* and *T. chebula*. These two dyes were selected for initial studies because they had greatest absorptions in the UVB region, shown in Table II. Secondly, these dyes have good affinities for cotton unlike most other natural dyes and can be dyed with as well as without use of mordants.

Detailed experiments were carried out using higher concentrations of these two dyes in combination with 3 mordants, alum, copper sulphate and ferrous sulphate. The undyed blank sample having a mean UPF of 4.91 was used as a standard for comparison. Results are reported in Table III. It can be seen that there is a dramatic increase in the UPF values for all samples when dyeing is carried out in combination with mordants. All samples offer good to excellent protection from UV radiation even at very pale shades. Maximum UPF values are obtained when dye concentration is between 9% and 12% owf. The SD values are low, implying uniform dyeing and reproducibility of results. The dyed and mordanted samples were also washed as per ISO 2 and the UPF values of washed samples were recorded. The values are reported in Table IV. It can be seen that after washing samples provide protection of a very high order implying that the protection is durable in nature.

Table IV UPF and K/S values of cotton samples dyed with natural dyes

Sample no	Dye	Conc. of dye (% owf)	Mordant	UPF	Std Dev.	Rated UPF	K/S	UPF after washing
1	QI	6	---	21.1	3.08	15	0.72	-
2	"	9	---	21.8	5.13	15	0.75	-
3	"	12	---	26.4	4.61	20	0.79	-
4	"	15	---	30.1	2.36	25	0.82	-
5	TC	6	---	16.1	0.25	15	0.76	-
6	"	9	---	20.6	1.04	15	0.95	-
7	"	12	---	22.1	1.19	20	1.02	-
8	"	15	---	24.5	0.89	20	1.06	-
9	QI	6	Alum	45.1	1.87	40	1.26	37.5
10	"	9	"	45.4	5.24	40	1.3	42.9
11	"	12	"	56.5	9	45	1.36	45.5
12	"	15	"	65.8	16.3	50	1.61	58.3
13	TC	6	"	35.5	0.778	30	2.2	33.2
14	"	9	"	43.8	13.5	40	2.59	41.5
15	"	12	"	49.3	3.7	45	2.98	46.1
16	"	15	"	52	3.92	45	3.52	50.9
17	QI	6	$CuSO_4$	36.3	3.1	30	1.5	31.9
18	"	9	"	36.7	0.31	35	1.7	33
19	"	12	"	39.5	1.6	35	1.88	37.8
20	"	15	"	40.2	1.8	35	1.92	39.1
21	TC	6	"	21.9	2.65	20	1.84	21.9
22	"	9	"	27.5	0.895	25	2.21	25.7
23	"	12	"	30.9	2.6	25	2.52	25.9
24	"	15	"	34.4	0.36	30	2.55	29.1
25	QI	6	$FeSO_4$	33.8	3.5	30	1.56	26.1
26	"	9	"	35.5	1.06	30	2.05	35.3
27	"	12	"	39	0.457	35	2.70	36.8
28	"	15	"	43.5	4.7	35	1.59	39.2
29	TC	6	"	26.9	4.62	25	1.69	23.3
30	"	9	"	28.1	2.6	25	2.12	27.4
31	"	12	"	35.1	2.8	30	2.21	31.2
32	"	15	"	49.9	4.8	45	3.38	39.4

K/S values of samples were also recorded. No relationship could however, be established between the depth of shade (K/S) and UV protection (UPF) provided by the dyed fabric. It is most likely that the chemical structure of the dye decides the UV absorption characteristics of dyes and therefore light shades can also impart good protection. Detailed studies on structure – property relationship can yield more information about this phenomenon.

Determination of antimicrobial activity of natural dyes on cotton fabric

The above study showed that Q. infectoria has good UV absorption properties. Hence it was decided to carry out studies on the antimicrobial activity of this dye against two bacteria, namely *P. vulgaris* and *E.coli*. Since most recipes for dyeing with this dye are based on the use of copper sulphate and ferrous sulphate as mordants for colour development, the effect of mordants and their concentration on antimicrobial activity was also studied. The commercial antimicrobial agent Fabshield was used as control.

Quantitative evaluation of antimicrobial activity on a solid substrate can be carried out suitably by colony counting. This method was employed to test the anti microbial efficacy of *Q. infectoria*. Results are listed in Table V and the effect of dye concentration on anti microbial activity is shown in Figure 1.

Table V Effect of dye concentration and mordant on antimicrobial activity of *Q Infectoria* on cotton fabric

Microbe	Sample type	Number of CFU $(x\ 10^5/ml)$	Reduction in CFU (%)
E.coli	Control	539	--
	Fabshield-treated fabric	20	96.6
	0.5% shade	545	0
	6% shade	466	13.9
	12 % shade	422	22.3
	15 % shade	14	93.5
	15% shade + 2% $CuSO_4$	348	33.9
	15% shade + 2% $FeSO_4$	512	4.8
P.vulgaris	Control	899	--
	Fabshield-treated fabric	14	94.3
	0.5 % shade	907	-0
	6 % shade	910	-0
	12% shade	885	1.4
	15 % shade	4	91.7
	15% shade + 2% $CuSO_4$	869	7.2
	15% shade + 2% $FeSO_4$	810	9.8

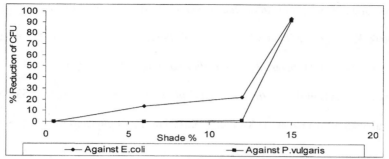

Figure 1 Antimicrobial efficacy of *Q. infectoria* against *E. coli* and *P. vlulgaris* on cotton fabric

It can be seen that the antimicrobial activity of fabric dyed with *Q. infectoria* increases with increase in the shade %. At 15% shade there is a sharp increase in activity with the fabric showing 97.4% reduction in CFU for *E.coli*. No activity is seen against *P.vulgaris* at less than 12% shade, while there is very high activity at 15% shade where the reduction in CFU is ~ 99.5%. The antimicrobial activity is better than that of standard antimicrobial agent – Fabshield.

An interesting observation was that the antimicrobial activity decreased significantly or was lost completely when mordant was used in combination with the dye. Loss in activity was found to be higher when $FeSO_4$ was used as a mordant. This could be because of the complex formation between dye and metal salts. The only functional groups present on these dyes are hydroxyl groups. Hydroxyl groups ortho to the chromophore are involved in co-ordinate bond formation with the mordant. If these same groups are responsible for the antimicrobial activity- when they get complexed no free groups are available to act on the microbes. Hence better is the complex formation, faster is the dye but lesser is the antimicrobial activity.

CONCLUSIONS

From the current study it could be established that natural dyes can indeed impart multifunctional properties to cotton fabric. Besides imparting fast and harmonious shades on cotton, *Quercus infectoria* can provide durable UV protection of a very high order. The fabric at the same time has a bactericidal effect and shows activity which is as good as that of commercial antimicrobial agents. Concentration of dye required is quite high being about 12 to 15% to make the fabric both UV protective and bactericidal. This is probably due to the low exhaustion of the dye on cotton. Mordants have a positive effect on UV absorption but a negative effect on antimicrobial activity. Further work needs to be done with natural dyes on different fibrous substrates to elucidate their protective properties.

REFERENCES

1 G K Munshi, Unpublished M.Tech Thesis, IIT Delhi, 2001.

2 I Ahmad and AZ Beg, *J of Ethnopharmacology*, 2001 **74** (2-3) 113-123.

3 G H Naik, K I Priyadarsini and J G Satav et al, *Phytochemistry,* 2003 **63** (1) 97-104.

4 L Kambizi and A J Afolayan, *J of Ethnopharmacology*, 2001 **77** (1), 5-9.

5 N Ivanovska and S Philipov, *Int J of Immunopharmacology*, 1996 **18** (10) 553-561.

6 M Hiramatsu, M Komatsu, Y Xu and Y Kasahara. *Pathophysiology*, 1998 **5** Suppl(1) 79.

7 T Takii, S Kawashima and Taku Chiba et al, *Intl Immunopharmacology*, 2003 3(2) 273-277.

8 S Chatterjee, S R Padwal Desai and P Thomas. *Food Research Intl*, 1999 **32** (7) 487-490.

9 A Mazumder and K Raghavan et al, *Biochemical Pharmacology*, 1995 **49** (8) 1165-1170.

10 Z Sui, R Salto, J Li, C Craik and P R Ortiz de Montellano, *Bioorganic & Medicinal Chemistry*, 1993 **1** (6) 415-422.

11 G Singh, OP Singh and S Maurya, *Progress in Crystal Growth and Characterization of Materials*, 2002 **45** (1/2) 75-81.

12 P Scartezzini and E Speroni, *J of Ethnopharmacology*, 2000 **71** (1/2) 23-43.

13 K S Babu, P V Srinivas, B Praveen, K Hara Kishore, U S Murty and J M Rao, *Phytochemistry,* 2003 **62** (2) 203-207.

14 Y Shinji; et al. US Patent Office, Patent Application No 20030082244, 2003.

15 M L Gulrajani and D Gupta, *Natural dyes and there application to textiles*, IIT Delhi, 1992.

16 Y Hitotsuyanagi, T Aihara and K Takeya. *Tetrahedron Letters*, 2000 **41**, (32) 6127-6130.

17 www.himalayahealthcare.com

18 A Redwane, H B Lazrek, *J of Ethnopharmacology*, 2002 **79** (2) 261-263.

19 M C Sabu and R Kuttan, *J of Ethnopharmacology,* 2002 81(2) 155-160.

20 A K Das, S C Mandal, S K Banerjee et al, *Jl of Ethnopharmacology*, 1999 **68** (1/3) 205-208.

21   M A Jafri, M Aslam, K   Javed and S Singh. *J of Ethnopharmacology*,   2000 70(31) 309-314.

22 T B Machado, AV Pinto et al., *Intl J of Antimicrobial agents*. 2003 **21** (3) 279-284.

23 D Prashanth, M K Asha and A Amit., *Fitoterapia*, 2001 72(2) 171-173.

Part V

Novel technologies

TANDEM WET-ON-WET FOAM APPLICATION OF BOTH CREASE-RESIST AND ANTISTATIC FINISHES

[1]John Pearson and [2]A. Elbadawi

[1]Department of Textiles, University of Huddersfield, Queensgate, Huddersfield, HD1 3DH, UK
[2]Industrial Research and Consultancy Centre, P O Box 268, Khartoum, Sudan

ABSTRACT

Wet-on-wet application of textile finishes is normally carried out using pad-mangles for application and stenters for drying and curing. The former method results in the release of significant quantities of effluent water, containing excess finish and auxiliaries. The latter method requires the consumption of large amounts of energy to remove the water. Overall, this results in a process which produces large amounts of waste and which uses large amounts of energy.

Foam application of finishes is known to minimise waste and to use much less thermal energy: the process is currently being used to apply crease-resist finishes. This paper describes a method for applying successive crease-resist and antistatic finishes using successive foam treatments without intermediate drying. The paper concentrates on the effects of using a range of dwell times of the foam on the fabric, which are critical to the success of the process.

Compared to the standard procedure, this technique would considerably reduce the amount of effluent produced and heat energy used. The usefulness of the process is also described in the paper in terms of the effectiveness of the finishes applied.

INTRODUCTION

Conventional pad-mangle processing

Addition of various finishes in one bath requires special chemicals and high skill and levels of expertise in recipe formulation and coupled with compatibility problems between the finishes, for instance when using an anti-static agent together with a crease-resist finish. The normal procedure is therefore to give an antistatic aftertreatment to the crease-resist finished fabric in a separate bath after the fabric has been cured. Traditionally, the pad-mangle technique has been used for each process, meaning that significant amounts of water are used as solvent and transporter of the finishes, and significant amounts of energy are used for subsequently removing that water. In addition, since not all of the solution of finish plus auxiliaries is fixed in the fabric, the pad-mangle process produces significant amounts of effluent, which has to be treated. The standard process, therefore, can be expensive and polluting.

Foam systems

Foam is an agglomeration of gaseous bubbles, usually of air, dispersed in a liquid and separated from each other by thin films of liquid, known as lamellae. The use of air to extend the volume of a concentrated chemical solution in the form of bubbles is known as a **foam system**. Foaming a liquor to a larger volume (usually between 5 and 50 times the liquor volume) allows convenient volumetric control and hence ensures even distribution of the liquor in the fabric. Foam systems could be the most logical alternative to the conventional pad-mangle system for application of finishes in tandem, since the reagents are applied to the fabric in the form of a foam, in contrast to the conventional process of impregnating the fabric with a dilute solution of the reagent. Since air replaces water as the transport medium for the reagents, substantial energy savings in the drying of fabrics, less waste disposal and enhanced quality of the product can be realised.

EXPERIMENTAL

Experimental work involved the tandem foam application of a crease-resist and an antistatic finish at varying dwell times (dwell time is defined as the time to elapse between the application of the two finishes). Four dwell times were used (30, 35, 40 and 45 seconds respectively) as being those most likely to be used in practice.

  The tandem wet-on-wet foam application method was that of Elbadawi and Pearson and was controlled to give a maximum wet pick-up up of 45% so that the potential advantages of foam add-on are not lost. This included crease-resist finish at 30% wet pick-up and antistatic finish at 15% wet pick-up. The crease-resist agent, which is an absorbed reactive finish, was applied before the antistatic agent, which is a topical finish.

  For comparison purposes, fabric samples were also finished using the conventional pad-mangle system at the 60% wet pick-up level used commercially:

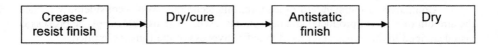

  Tables 1a and 1b show the recipes for the crease-resist and antistatic finishes respectively; table 2 shows the fabric specification. Warcosoft NI 150 softener was included in the anti-static finish recipe only, since it was found logical to incorporate the softener (which aims to improve handle, in part a surface property) with the topical finish to operate at the fabric surface.

Table 1a Crease-resist finish recipe

reagent	Concentration (gl⁻¹)	
	Pad-mangle	Foam
Permafresh[a]	55	110
Catalyst 8316	22	44
Laviron 118S[b]	-	15
Kelzan S[c]	-	2
Water	Adjust to 1 l	Adjust to 1 l

196

Table 1b Antistatic finish recipe

reagent	Concentration (gl⁻¹)	
	Pad-mangle	Foam
Lurotex A25[d]	35	140
Warcosoft NI 150[e]	20	80
Laviron 118S[b]	-	15
Kelzan S[c]	-	2
Water	Adjust to 1 l	Adjust to 1 l

Notes to tables 1a and 1b:

a: Permafresh	DMDHEU-based crease-resist finish (Contract Chemicals)
b: Laviron 118S	amphoteric, fatty amine oxide surfactant (Henkel)
c: Kelzan S	xanthan gum thickener (Kelco International)
d: Lurotex A25	Non-ionic polyamide anti-static agent (BASF)
e: Warcosoft NI 150	fatty glyceride blend non-ionic softener (Contract Chemicals)

Table 2 Parameters of fabric used in this work

fabric composition	50% polyester/50% cotton
weave structure	satin
mass/unit area	189 gm^{-2}
thickness	0.45 mm

Foamed finishes were prepared using a Mondomix foam generator and were applied on the knife-over-roller using a Werner-Mathis coating device, set to give a coating gap of 0.3 mm. The coated fabrics were passed between a pair of rollers to collapse the foam. Fabric through speed was 4 m min^{-1}; the roller nip pressure was 1 kg cm^{-1}. Fabric samples were dried/fixed at 150°C for 3 min. Foam densities and coating gap were determined to give the correct wet pick-up and solid add-on values. Roller nip pressure and speed, and temperature and time were the standard values for the two finishes.

RESULTS

The economic and environmental advantages of foam processing are well established. However, these advantages would not count should the effectiveness of the finishes applied by using foam systems be significantly worse than by using conventional application methods. The finished fabrics were therefore tested for abrasion resistance, crease recovery angle, tear strength, tensile strength and fabric breaking extension, fabric shrinkage and de-cling time. The results are shown in table 4.

Table 4. Mean values of fabric parameters resulting from the use of foam and conventional systems for applying finishes in tandem

Mean values	Un-finished	Pad-mangle finish	Dwell time (sec) – foam finished			
			30	35	40	45
abrasion resistance (rubs x 1000)	29	16	25	21	16	19
warp crease recovery angle (°)	100	123	115	117	117	118
weft crease recovery angle (°)	96	131	125	122	123	121
warp tear strength (N)	14.2	11.6	10.7	10.2	10.6	10.4
weft tear strength (N)	12.3	10.8	9.7	9.6	9.4	9.4
warp tensile strength (N)	496	403	386	399	398	401
weft tensile strength (N)	470	382	374	377	367	361
fabric breaking extension (weft %)	14.2	11.6	11.5	11.5	11.4	11.4
fabric breaking extension (warp %)	21.7	17.7	17.4	17.4	17.0	17.2
warp fabric shrinkage (%)	1.7	0.2	1.1	0.7	0.0	0.0
weft fabric shrinkage (%)	0.0	0.0	0.0	0.0	0.0	0.0
decling time (sec)	19.7	17.5	14.5	13.7	12.5	11.7

DISCUSSION

Abrasion resistance

For foam-finished samples, abrasion resistance has a tendency to decrease with an increase in dwell time. This may be because the longer the dwell time, the greater the chance that the reactant will have to cross-link with the fibres before applying the anti-static finish, minimising any interaction between the two finishes. Excessive cross-linking may have occurred, leading to increased rigidity or decreased flexibility of the fabric, reducing the abrasion resistance. Dwell time had no significant effect on abrasion resistance.

Crease recovery angle

Both application methods are seen to increase the crease recovery angle, conventional finishing giving slightly better results. This could be because foam processing may lead to uneven distribution and effectiveness of cross-linking. Dwell time is seen to have no significant effect on crease recovery angle of the foam finished samples. Weft crease recovery is seen to be higher than that in the warp direction.

Tear strength

Both application methods reduce fabric tear strength, and foam processing was worse than conventional processing in this respect, probably due to uneven distribution of cross-linking weakening the fabric. Warp direction tear strength was higher than that in the weft direction for all finished fabrics, due to the normal higher strength of warp yarns in a fabric. Dwell time had no significant effect on tear resistance.

Tensile strength

Both application methods reduced the tear strength of the finished fabric to a similar degree. Dwell time had no significant effect on warp or weft tensile strength.

Fabric breaking extension

Both application methods reduced the breaking extension of the finished fabrics to a similar degree. Again, dwell time had no significant effect on this parameter. Breaking extension was slightly higher in the weft direction than in the warp direction, probably due to lower weaving tension and twist level in the weft.

Fabric shrinkage

Both finishing treatments decreased fabric shrinkage: there was no shrinkage in the weft direction in any of the treated fabrics, but there was a small amount of shrinkage in the warp direction. Higher dwell times appear to increase the extent of cross-linking, which could reduce subsequent fabric shrinkage.

Decling time

Both application methods improved the anti-static properties of the finished fabric, in this case to a greater extent when using the foam system. This is probably due to the fact that the foam system concentrates the finish to the fabric surface where it is most effective. The decling time decreased as the dwell time was increased, possibly due to there being a greater time available for surface application to dissipate in the case of a longer dwell time, increasing the concentration of the antistatic finish on the fabric surface.

CONCLUSION

The physical performance of the tandem wet-on-wet foam finished fabric was comparable to that of the conventionally finished fabric, although the former method gave a slightly decreased level of performance in some areas. This suggests that there is a future opportunities for improving the performance of the foam system *via* better control of foam properties and foam processing parameters.

More work needs to be done on the effect of dwell time on fabric performance. Shorter dwell times (below 30 s) will allow for rapid, commercially acceptable tandem wet-on-wet processing if successful in producing fabrics with acceptable levels of performance.

AN ALL NATURAL SLIP RESISTANT & ABSORBENT FIBROUS MATERIAL

Sebastian R. Hutchinson

Departments of Textile Engineering, Chemistry & Science and Textile & Apparel Technology & Management, College of Textiles, North Carolina State University, 8301 Research Drive, Raleigh, NC, 27695, USA

Keywords: coefficient of friction, moisture, absorbent, sweat loss, exercise, yoga

ABSTRACT

PVC-based exercise mats are the de facto standard practice surface for the millions of regular American *hatha* yoga practitioners. The 'sticky mat' provides a lightweight slip-resistant surface in dry conditions. However, this mat has inherent performance drawbacks and longstanding biocompatibility concerns.

'Postures' or *asana*-styles demand different properties from the standard exercise mat. The amount of perspiration generated by practitioners varies by style and athletic ability. In both cases the individual requires a slip-resistant practice surface. Standard foamed closed-cell PVC mats and cotton rugs provide insufficient properties in wet and dry conditions.

This study investigates changes in the ability of a mat to resist slippage upon absorption of water for the PVC mat, cotton rug and a novel slip-resistant and absorbent fibrous material. Dynamic coefficients of friction as a function of absorbed water are examined. The efficacy of each mat to resist slippage with absorbed water is correlated to expected sweat generation during practice. A prior predictive sweat loss response model to metabolic rate, environment and clothing is used.

INTRODUCTION

Polyvinyl chloride (PVC) is a soft material lauded for its inertness and subject to longstanding biocompatibility and environmental concerns. Various plasticizing agents used to modify the polymer behavior of the inexpensive material are also considered questionable. Yet today PVC petrochemicals are the second largest class of thermoplastics spanning the consumer marketplace from plumbing to children's toys.[1-4]

Hatha yoga

PVC found its way into the exercise and sporting goods industry as yoga mats in the 1970s. Hatha yoga 'postures', or *asanas*, comprise a subset of one of the six paths of Indian yoga ranging from devotion (*bhakti*) to knowledge (*jnana*). The word yoga itself translates loosely as 'yoke' or 'union,' describing ways to integrate the physical, mental and spiritual aspects of human existence. This system was most recently brought to the west in the 1950s and 1960s. Subsequently, *hatha* yoga has diverged into many styles catering to many types of practitioners. As of 2002, an estimated 18 million Americans practice some form of *asanas*, or the misnomer 'yoga'.[4]

In India, *asanas* are traditionally practiced on a thin cotton rug or a dirt floor. Yet in the west, practitioners found this material an insufficient means of providing traction that maximizes grip while preventing slippage.[5]

Styles & practice conditions

Numerous styles of *asanas* have emerged since the general yoga diaspora in the mid 20th century. Table 1 outlines popular types associated by physical difficulty. Students experience a range of practice conditions. More vigorous styles create substantial amounts of perspiration. Gentle to moderate difficulties tend to generate less. The amount and rate of perspiration varies by individual, instructor, and style. A typical session lasts for 30 minutes to 2.5 hours and can generate over 1 liter of perspiration.

For those who do not sweat, a synthetic sticky mat gives sufficient traction. For those who perspire even slightly, it offers little to no grip. For vigorous styles, perspiration can be moderate and this group uses cotton rugs. However, when wet the rug yields limited traction and does not resist slipping on smooth floor surfaces.

Table 1. Popular styles of western Hatha yoga[6]

Style	Description
Astanga	Vigorous
Bikram	Moderate, hot
Kundalini	Gentle
Iyengar	Moderate
Sivananda	Moderate
Viniyoga	Gentle
Vinyasa	Vigorous

PVC

Raw, PVC is a hard material that softens and shrinks at relatively low temperatures. A range of degradation mechanisms including thermal, chemical, photolysis, and irradiation yield a variety of often harmful byproducts.[7-12] Most notable are hydrogen chloride and chlorinated benzenes, notably dioxins.[3,13] To alter the behavior of the material, plasticizers such as di(2-ethylhexyl) phthalate (DEHP), diisononyl phthalate (DINP) and heavy metal compounds (lead, cadmium, mercury, zinc, tin & barium) have been used to stabilize and yield a variety of behaviors.[14] Plasticizing agents and products of degradation fuel the regulations for health and environmental concerns surrounding the manufacture, consumption and disposal of plasticized PVC.[15-17]

Regulations

The Delaney Clause of the U.S. Food, Drug and Cosmetic Act of 1958 states, "no additive shall be deemed to be safe if it is found…to induce cancer in man or animal…" Not until angiosarcoma (liver cancer) and what was later labeled the 'PVC Disease' were traced to autoclave vinyl chloride workers in the early 1970s did the U.S. Food and Drugs Administration regulate the amount of human contact.[12] A proposed regulation issued in

September 1975 permitted continued use of PVC in contact with foodstuffs, "where the potential for migration of vinyl chloride is diminished to the extent that it may not reasonably be expected to become a component of food".[18]

Table 2. Oekotex 100 Certified, chemical limits of sticky mats by Simply Yoga, London, UK[19]

Attribute	Value
pH	4.0-7.5
Formaldehyde	75ppm
Extractable heavy metals	92ppm
(arsenic, lead, cadmium, chromium, etc.)	
Mercury	0.02ppm
Pesticides	1ppm
(DDT, Lindane, Hexachlorobenzene)	
Phenols	2ppm
Organic tins	1ppm
Chlorobenzenes and chlorotoluenes	1ppm
Biocides	*None*
Forbidden flame retardants	*None*

Controversies

One of the more recent controversies surrounding the use of plasticized PVC in consumer products came with DINP in children's toys. Of particular interest were teething rings imported from China, which contained 40-50% by weight. In early 1997, a Danish ban led to efforts from various EU countries to control or limit toys containing DINP. Greenpeace's 'Exeter Report' of 1997 re-initiated worldwide controversy over the hazards of PVC and compositions plasticized with DINP. After a series of tests by Health Canada, Greenpeace and the National Environmental Trust petitioned the US Consumer Product Safety Commission (CPSC) to ban the material in related applications. Yet effects of low-level exposures for short durations may not induce acute symptoms.[14] As a result, rather than banning, the CPSC issued a request for manufacturers to discontinue production of consumer products containing DINP. In the EU, however, a recent directive phases out or controls the use of DINP in many plasticized PVC toys.

Worldwide production

Despite the controversies, PVC growth continues. The largest growth is shown in developing Asian countries where lack of social awareness and environmental regulations enable its unfettered expansion. In 2002, expected North American production of PVC was 9,350,000 metric tons with a slowing average annual growth rate of 4%. Compared with Asia, 12,920,000 metric tons with a 12% growth rate had doubled capacity in 5 years. In general, usage of PVC continues where lightweight, durable, and economic materials are needed.[2,20]

At single high-dose exposures, many PVC plasticizing and filler chemicals are known or suspected carcinogens. Reactions from chronic low-level exposures are not as well understood.[14] Before an assertion is made about the toxicity of PVC sticky mats, accurately identifying the contents is critical. Table 2 shows tested levels of toxic chemicals in an Oekotek 100-certified sticky mat by Simply Yoga, London, UK. Phthalate and other plasticizers are not included in this test. Toxic contents of other commercial sticky mats are not publicly available. In fact, Hugger Mugger (Salt Lake City, UT, USA), marketer of the most popular PVC 'sticky' mat the Tapas® Mat, makes no public disclosure about the contents, other than it is PVC-based.

Table 3. Absorptive capacity and tested amount of water by mat type

Mat	Mass Cotton (g)	Capacity (mL)	Tested (mL)
Tapas® mat	0	1	0 1
Yoke Mat™	30	100	0 6 12 25 50 100
Cotton rug	65	200	0 50 100 200

MATERIALS & METHODS

Three yoga and exercise mats were used in this experiment: a Tapas® Mat by Hugger Mugger, a Yoke Mat™ by Complete Circle (P.O. Box 1286, Fuquay-Varina, NC, 27526, USA), and a cotton rug by Prana (Vista, CA, USA). The Tapas® mat was composed of plasticized closed-cell foamed PVC. The cotton rug is a weft-faced plain weave cotton rug. The Yoke Mat™ is a proprietary all-natural fibrous material.

A 16.5 cm X 61.0 cm piece of each mat was cut and soaked in tap water for 3 hours to establish the maximum absorptive capacity. Mats were tested for slip force at a range of absorbed water from 0 mL to maximum capacity. Table 3 summarizes the capacities and tested amounts.

After the capacity for each mat was established, the mats were allowed to dry for 2 days. The predetermined test water level was evenly applied and allowed to condition for 15 minutes. Once the water was absorbed, the mat was placed on a smooth enamel metallic surface. This simulates the typical wood, ceramic or linoleum floor surface. Concrete and clay bricks were used to apply a specific load over a given area, see Fig. 1. Weight applied varied from approximately 1.6 to 16.8kg. A calibrated spring balance by

Chantillon's of New York, ca. 1892 was used to measure the amount of force required to induce mat slippage, or slip force (F_S), see Fig. 2. Once all measurable loads for the given absorbance were completed, water was again evenly added such that the total water added equaled the next test level. This iterative cycle was completed once the total capacity was reached.

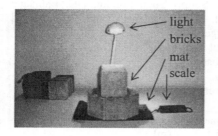

Figure 1. Slip force Measurement Setup

Figure 2.
Scale

RESULTS

Linear models fit well for each slip force by load curve. The slope is effectively the dynamic coefficient of friction. Plots of F_S by load show different trends for each mat, Figs. 3-5. Table 3 shows slope coefficient, y-intercept and r-square coefficient of determination. Structural and surface changes with absorbed water as well as experimental variation account for the nonzero and negative intercepts. Analyses indicate strong correlations.

The slip force for the Tapas® Mat significantly decreased with small applications of water. The mat itself tears when pulling under high loads. Since the mat absorbs no surface water, a film is produced with application of 1mL of water. Additional water yielded no further reduction in slip resistance. The slip resistance of the cotton rug increased to an observed maximum at 100mL water. Both 50mL and 100mL showed similar curve trends. At the 200mL level, the slope decreased, albeit 70% higher than when dry.

The slip force for the Yoke Mat™ shows a wider range of response depending on the load and level of water. Up to 25mL, small levels of absorbed water show a higher resistance to slippage than wetted PVC. The slope decreases with larger amounts of water.

Each mat shows a similar threshold slip force of approximately 70 N at the maximum load of 770 kg·m⁻². Fig. 6 shows slip forces for all mats at and above 70 N. For the Tapas® Mat, the threshold force is a minimum. For the cotton rug, it is a maximum. The Yoke Mat™ shows an intermediate threshold at 70 N.

The results of the linear fits for the slip force versus load curves were plotted against the water-to-cotton mass ratio. Figs. 7-9 show the plots of F_S-by-load linear-fit coefficient by mass ratio. For the Tapas® Mat and Yoke Mat™, the coefficient of friction decreased steadily with more absorbed water. The Yoke Mat™ showed a larger range of effective slip force resistance. The rug showed a peak force just under 2:1 mass ratio.

Figure 3. Slip Force by Load, Tapas® Mat

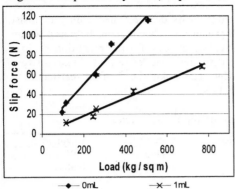

Figure 4. Slip Force by Load, cotton rug

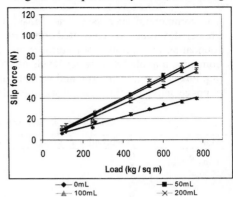

Figure 5. Slip Force by Load, Yoke Mat™

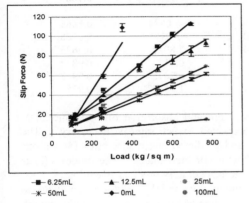

Figure 6. Threshold Slip Forces by Load

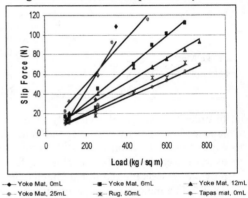

Figure 7. Tapas Coeff Fit

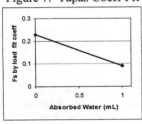

Figure 8. Yoke Coeff Fit

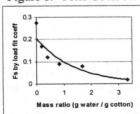

Figure 9. Rug Coeff Fit

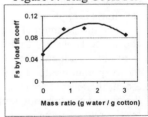

DISCUSSION

The commercial success of the Tapas® Mat is due to its demonstrated ability to resist greater slip forces than a wetted cotton rug. This property is advantageous both as a secure foundation on smooth floor surfaces and by providing traction for the practitioner. In a comparison at absorbed water levels of 25% capacity, the basic Yoke Mat™ resists larger slip forces than a cotton rug and wetted Tapas® Mat. The Yoke Mat™ offers resistances to slippage above the threshold at water levels up to 12%.

Table 4. Linear correlations of slip force by load

Mat	Absorbed water (mL)	Coefficient	Y-intercept	R-square
Tapas® Mat	0	0.2293	3.8840	0.9692
	1	0.0899	0.3196	0.9867
Cotton rug	0	0.0499	1.8380	0.9870
	50	0.0971	-0.7411	0.9897
	100	0.0982	0.3191	0.9753
	200	0.0845	0.1391	0.9885
Yoke Mat™	0	0.2745	-8.8784	0.8465
	6	0.1680	-3.0191	0.9788
	12	0.1204	2.8027	0.9625
	25	0.0886	0.4124	0.9877
	50	0.0787	0.2672	0.9948
	100	0.0174	1.1475	0.9789

In order to evaluate the viability of the slip-resistant and absorptive fibrous material in practice, an assessment of conditions and expected sweat rate is necessary. The experiments of Shapiro, et al[21,22] show actual and predicted sweat loss rates for various exercises, environments and clothing for 34 heat-acclimated males. The model adjusts previous models for additional interactions and more accurately predicts sweat loss.

While a majority of yoga practitioners are female, this model provides a framework to extrapolate sweat loss. The study correlates such human parameters as age, weight, height, body surface and body fat with environment conditions of temperature, relative humidity and convection with thermal insulative values of clothing ensembles. Exercise conditions consisted of walking at a speed of 1.34 m·s^{-1}. For the purpose of yoga, the clothing value for shorts is applicable.

The environmental and exercise conditions closely representing the styles of yoga is shown in Table 5. Sweat loss is approximate and adjusted as amounts depend on the individual's athletic ability. The amount of sweat reaching the mat is further dependent on the evaporation rate. In order to compare the actual values for the absorbed water levels in a mat, the sweat loss must be taken in proportion of the body surface area in contact with the mat. This does not apply for heavy perspiration rates where drips are formed. At these levels, rate of transfer of moisture to the mat increases, as does the amount absorbed by clothing. For vigorous conditions, the sweat loss is decreased by the surface area covered by shorts, which compensates for body area whose sweat is absorbed by the shorts. The surface area of mid-thigh-length shorts is 12.01%.

Based on the size of an average western female body, the surface area of hand and foot is 0.009677m² and 0.01742m². The total surface area is 0.05419 or approximately 5.42% of total body surface area. Using this ratio, the amount of expected moisture directly transferred to the practice surface is calculated as a fraction of the non-evaporated sweat loss. In the studied form, the Yoke Mat™ more effectively resists slipping than the PCV mat and cotton rug at water levels above 12.5mL. This corresponds to an absorption rate of 125 mL·m⁻², which is range a large portion of practice styles and athletic abilities.

Table 5. Sweat rate[†] by style and moisture absorption

Asana Style	Description	Non-evaporated Sweat loss $(g \cdot m^{-2} \cdot h^{-1})$	Moisture Absorption for Practice Surface $(mL \cdot m^2)$
Astanga	Vigorous	25 – 560	6.0 – 151.8
Bikram	Moderate, hot	580 – 932	157.2 – 252.6
Kundalini	Gentle	0 – 50	0 – 13.5
Iyengar	Moderate	0 – 126	0 – 34.1
Sivananda	Moderate	0 – 126	0 – 34.1
Viniyoga	Gentle	0 – 50	0 – 13.5
Vinyasa	Vigorous	25 – 560	6.0 – 151.8

[†] modified from *Shapiro, et al*[21,22]

CONCLUSIONS

The technical applicability of the Tapas® Mat is demonstrated in Fig. 6 for the adjusted sweat loss rates during a one hour course. The slightly-moist conditions of many practitioners require a slip force above the threshold. Yet, both the PVC mat and cotton rug fail to provide adequate slip resistance. Most practice in slight to moderate levels of absorbed perspiration, and the Yoke Mat™ provides an effective and all natural solution. Vigorous routines of seasoned practitioners may also fall into an effective range. Material selection and fabrication specifications will address this deficiency and may significantly improve the slip force at lower loads and higher water levels.

In general, the health and performance motivations for a machine-washable and dryable all-natural slip-resistant and absorbent fibrous material are shown. Effective coefficients of friction for three mats at different levels of absorbed water define the performance areas. The Yoke Mat™ offers resistance to slippage for most *asana* styles.

FUTURE WORK

The design of the Yoke Mat™ was not optimized for structural and surface changes due to swelling. High performance materials may also be chosen to enhance the overall properties of the material. The fibrous material may find other consumer and industrial applications where slip resistance and water absorption are desired.

ACKNOWLEDGEMENTS

The author extends his gratitude to Chris Moses of the Institute of Textile Technology for editing this manuscript, Beryl Bender Birch and Stan Woodman of The Hard & The Soft Astanga Yoga Institute for their *asana* styles discussions, Prana and Forrest Yoga Circle for support, and the International Sivananda Yoga Vedanta Centres of Woodbourne and NYC, NY for their superb intensive 6-month training in *asanas, karma* yoga and Vedanta.

REFERENCES

1 J A Kent, *Riegel's Handbook of Industrial Chemistry*, 10th Edition, New York, Klewer Academic / Plenum Publishers, 2003.

2 J A Tickner, 'Trends in world PVC industry expansion: a Greenpeace White Paper', http://www.ecologycenter.org/iptf/plastic_types/TrendsinWorldPVC(GP).htm, June 19, 1998.

3 J Wypych and A D Jenkins, Editor, Polyvinyl *Chloride Stabilization, Polymer Science Library 4,* New York, Elsevier Science Publishing Company Inc., 1986.

4 S Hutchinson, 'The Sticky in Your Mat,' *Yogi Times*, April 2004, 18.

5 C Kleiner, 'Mind-body fitness,' *U.S. News & World Report*, 2002 May 13 **132**(16).

6 F P Ruiz, 'Sticky Business', *Yoga J*, Winter 2000-2001.

7 J Cook, 'Not All Yoga is Created Equal', *Yoga J,* Winter 1999-2000.

8 C Winder, 'The Toxicology of Chlorine', *Environmental Research Section A*, 2001 **85** 105-114.

9 K A Graeme and C V Pollack, Jr., 'Heavy Metal Toxicity, Part I: Arsenic and Mercury', *J Emergency Medicine*, 1998 **16**(1) 45-56.

10 K A Graeme and C V Pollack, Jr., 'Heavy Metal Toxicity, Part II: Lead and Metal Fume Fever', *J Emergency Medicine*, 1998 **16**(2) 171-177.

11 D Kriebel, 'The Dioxins: Toxic and Still Troublesome', *Environment,* 1981 **23**(1) 6-3.

12 P F Infante, 'Observations of the Site-Specific Carcinogenicity of Vinyl Chloride to Humans', *Environmental Health Perspectives*, 1981 **41** 89-94.

13 W K Lelbach, 'A 25-Year Follow-up Study of the Heavily Vinyl Chloride Exposed Workers in Germany', *American J of Industrial Medicine*, 1996 **29** 446-458.

14 E D Owen, Editor, *Degradation and Stabilisation of PVC*, London, Elsevier Applied Science Publishers, Ltd, 1984.

15 C F Wilkinson and J C Lamb IV, 'The Potential Health Effects of Phthalate Esters in Children's Toys: A Review and Risk Assessment', *Regulatory Toxicology and Pharmacology*, 1999 **30** 140-155.

16 R H Burgess, Editor, *Manufacture and Processing of PVC*, New York, Macmillan Publishing Co, Inc, 1982.

17 F P La Mantia, Editor, *Recycling of PVC & Mixed Plastic Waste*. Ontario, ChemTec Publishing, 1996.

18 A Whelan, and J L Craft, Editors, *Developments in PVC Production and Processing – 1*, London, Applied Science Publishers, Ltd., 1977.

19 J V Koleske and L H Wartman, *Poly(Vinyl Chloride)*, New York, Gordon and Breach Science Publishers, 1969.

20 Anon, 'Sticky Yoga mats – Q and A', Simply Yoga, www.simply-yoga.co.uk, article 17/120504.

21 G Gappert, J F Coates and I Leveson, *Vinyl 2020: Progress, Challenges, Prospects for the Next Quarter Century*, Morristown, NJ, The Vinyl Institute, 1996.

22 Y Shapiro, K B Pandolf, and R F Goldman, 'Predicting Sweat Loss Response to Exercise, Environment and Clothing', *European J of Applied Physiology*, 1982 **48** 83-96.

23 Y Shapiro, D Moran, Y Epstein, et al, 'Validation and adjustment of the mathematical prediction model for human sweat rate responses to outdoor environmental conditions', *Ergonomics*, 1995 **38**(5) 981-986.

LASER AND DEVORE

Janet Stoyel
University of the West of England, Bristol.

ABSTRACT

This paper will discuss the printing process popularly known as devore and will highlight the problems associated with the disposal of residual dyes, chemicals and waste products, generated as possible pollutants to the water course, by this process. A comparison of conventional printed devore will be made with an environmentally responsible futuristic devore effect achieved through laser technology.

Keywords: Laser, Devore, environmental, substrates.

THE IMPETUS

Environmental legislation and the attendant regulatory pressures pertaining to waste management are felt throughout the world of textile production.

It is a recognisable fact that the world of fashion is not a knowledge-based industry, and, in general the contemporary fashion designer is less concerned, often unaware, of the negative effects to the environment created by their design decisions, the problems associated with textile production are largely ignored the philosophy being: 'after all it is someone else's problem'.

This attitude begins, and is often fostered, through the print and design training that U.K. students receive at college and university. Within the majority of these courses there is no attention given to the disposal of residual dyes or chemicals that remain after a printing process. It is to be noted that in general, interdisciplinary textile cross-fertilisation within the textile disciplines at university level, although much desired, is more often than not impossible to achieve. It is not surprising therefore that graduating fashion designers have no concept of textile production or how the subsequent realisation of the desirable fabrics they order for garment collections happens.

DEVORE

There is a saying prevalent within print education: 'if you cannot do it... devore it'. Good print **is** difficult to master and achieve, the suggestion being: devore always looks goods, even as a simple stripe and spectacular effects are possible when combining multiple design solutions. But, the print process commonly known as "Devore" or "Burn-Out" is a particularly renowned pollutant, not only does the process generate chemical and colour waste it also generate large quantities of fibrous sludge.

The devore process works by chemically destroying one fibre or more in a bi-component or multi-component substrate, generating effects that resemble lace and creating textiles with different degrees of translucency, transparency and opacity. It is a misnomer to think this process is a 'cheap and cheerful' alternative to conventional print, it is not, the chemicals and dyes employed for manufacture combined with the energy hungry processes reliant upon steam, dry, bake and repeated wash-off combined with the labour intensive printing action ensures a high price end product.

All print departments within educational establishments have their own set of recipes for student application. The following two recipes for devore are typical of such print related handouts.

DEVORE PRINTING RECIPES.

RECIPE 1: DEVORE OR BURN-OUT PRINTING

Fabrics: Silk/viscose; polyester/cotton; nylon/viscose.

Devore or burn-out printing produces a 'Lace' effect of transparent/translucent areas and opaque areas. It relies upon one of the fibres being degraded by components in the print paste leaving the other fibre as an open network. Experimentation is needed to achieve good results.

Recipe:

500	grams	Thickening 301
20	grams	Glycerine
150	grams	Aluminium sulphate
40	grams	Tartaric acid
290	grams	Water
1000	grams	Litre of print paste.

Method:

1. Weigh out thickening 301, water and glycerine into a plastic pot.
2. Weigh out the aluminium sulphate and tartaric acid. Add to the plastic pot and mix well on a high-speed mixer.

PRINT .... DRY .... BAKE .... RUB-OUT .... WASH .... DRY

Printing:

Make sure that the print paste penetrates the fabric well.
For velvet fabrics print on the back of the cloth not on the pile. This needs to be taken into consideration when making screens, as any design will come out back to front. The number of pulls you will require will vary depending upon your strength, and the type of screen used, but you will need at least six pulls.

Baking:

Bake at 170 °C for 5 minutes until the cellulose turns brown/black in colour.

Rub-out:

Rub-out as much of the burnt-out cellulose as possible before washing the fabric.

It is important that the a vacuum cleaner is used to suck up the loose fibres and a face mask and gloves are worn for protection.

Washing off:

1. Wash in warm water with detergent until clear of all burnt areas.
2. Rinse well in warm water.

RECIPE 2: DEVORE OR BURN-OUT PRINTING

Fabrics: Wool blends; wool/polyester; wool/cotton.

Devore or burn-out printing produces a 'Lace' effect of transparent/translucent areas and opaque areas.

It relies upon the wool fibres being degraded by the sodium hydroxide in the print paste leaving the other fibre as an open network. Experimentation is needed to achieve good results.

Recipe:

750   grams Solvitose thickener
50   grams Sodium hydroxide Pellets
200   grams Cold water
1000 grams Litre of print paste

Method:

It is necessary to wear full protective clothing when preparing, printing and washing off: goggles, rubber gloves, long sleeved overall.

1. Very carefully weigh out the sodium hydroxide pellets and add them very slowly to the cold water in the plastic pot. This process produces a large amount of heat so allow too cool in a safe place. Label the container clearly to avoid any accidents.

2. Weigh out Solvitose thickening into plastic pot.

3. Very carefully and slowly add the cold Sodium Hydroxide to the Solvitose, mixing well.

4. Label the container with the correct hazard warning label and put in a safe place until it is used for printing.

PRINT ...... DRY ...... STEAM ...... WASH-OFF ...... DRY

Printing:

Wear goggles, gloves and a long sleeved overall. Make sure that the print paste penetrates the fabric well. The number of pulls you will require will vary depending upon your strength and the type of screen used: but you will need at least six pulls.

Wash screen. Very carefully label areas of printing to warn others of the dangers of caustic burns while the print dries.

No mention is made of eventual waste disposal and experience proves that the number of print students who adhere to such health and safety guidelines suggested within these hand-outs is negligible.

The textile practitioner

Janet Stoyel is a practicing artist and practitioner in the applied arts, she owns and operates a small, successful decorative textile business, established in 1994, The Cloth Clinic, and has a fractional post as a Senior Research Fellow at the University of West of England, Bristol.

Personal ethics

As a research practitioner investigating: Fabric Finishes and Treatments for Apparel and Interior Applications, whilst studying for a Master of Philosophy degree, at the Royal College of Art, London, in 1992-94, environmental concerns arose, regarding the safe disposal of textile generated waste in the form of chemicals and dye effluent. Subsequent research showed it is practically impossible to remove **ALL** such potentially polluting substances from the water chain. This fact prompted in depth research and the hunt for a technology that would preclude the use of dyes, chemicals or wet processes for the creation of permanent patterning and decorative effects, and, which would perform in what was deemed to be an environmentally responsible manner. A search for a technology which would be ecologically sound and energy efficient, which would be capable of creating fully-recyclable textiles without and extraneous fibre, yarns, threads, equipment and tools to eventually become recognised pieces of acceptable textile machinery.

It became a matter of personal concern, whilst researching textile finishing, printing in particular, the sheer quantities and volumes of textile waste and effluent generated by student activity. Research has shown that government legislation is firmly in place for the disposal of industrial textile wastes and effluents, but, small non-commercial establishments with lower quantities of such matter are not regulated and as such almost all effluent and waste is released directly into the drains, sewage system and water courses.

If a university has a three-year course with twenty print students per year printing, a considerable amount of waste is released slowly into the waterways system and so manages to avoid legislation and controls for waste. Magnify this number of students in an area such as Greater London with its plethora of colleges and universities offering print courses, include the graduate textile printers who establish workshops in the surrounding area and the levels of textile effluents rises considerably, and, as such may even rival industrial output and be worthy of legislatory action.

The fact that the devore print process produces so much damaging effluent and yet is considered to be so very beautiful, desirable and costly when applied to fashion items is a complete paradox and was the final deciding factor in the search for a new clean and economical technology.

Lasers

Laser is acronym for: Light Amplification by Simulated Emission of Radiation.

Lasers are not new and have been used in many applications, including the Ministry of Defence for many years. In this instance what is new is the application to which they are being directed. Working in close collaboration with a UK-based company who manufacture laser optics for a missile firing system on the Challenger tank, and after intense sampling of numerous base substrates over a two year period, a personally designed and commissioned laser system was subsequently developed and installed in the workshop premises, of The Cloth Clinic, in the Blackdown Hills of East Devonshire.

Substrates

Exhaustive sampling to determine the perfect mix of synthetic components with which to create the ultimate effects required for commercial viable, comfortable, wash-and-wear product, resulted in engineered-base substrates combining intimate blends of polyester and polyamide being manufactured by a collaborating French weave concern. To dispel myths and misconceptions pertaining to polymeric exclusivity for laser processing, it must be noted that base substrates other than thermoplastic **are** suitable for laser processing, the final decision regarding the type of base substrate, its fibre composition and integral structure are reliant upon end use application.

The laser is capable of producing an effect very similar to devore effects whereby part of the surface of a substrate is abraded by a single or multiple high-speed laser heads, leaving a transparency and translucency similar to the print effects of chemical degradation by devore. The laser 'burn-away' effect is practically instantaneous, different levels and depths of erosion are feasible and detailed three-dimensional effects may be easily achieved. Designs are generated via digital camera, subsequently grey-scale techniques are readily achievable. The laser process generates minimal dust; this dust is extracted through a fume extraction system complete with air filter guards. The equipment is computer controlled and includes a comprehensive self-registration system to ensure perfect accuracy. The laser is a low energy, 10 watt, sealed unit and works from a single-phase electricity supply. Running cost is minimal balancing, out the expensive initial outlay for capital equipment acquisition. Many different types of substrate may be processed, substrates as sheer as polyester chiffon and as coarse as corduroy have processed successfully.

Product

The textile product from conventional devore, created by the print-based process, is soft and fluid and it often incorporates dyed and printed pattern. A silk and viscose velvet substrate is the most popular base material for luxurious fashion garments and scarf accessories. The finished articles require careful handling and need dry-cleaning. The print process is also popular for furnishing fabrics on cotton or linen mixed fibre substrates when the finished article is used to create room dividers, curtain and window treatments, and such end uses maximise the transparency/opacity characteristics of the process.

Laser devore on the other hand may consist of any type of substrate base. It is not necessary to create mixed fibre blends to 'burn-away' one or more of the fibre components, since the sensitivity of the technology allows partial degradation of the textile surface to be achieved. Handle and drape characteristics are generally reliant upon the base substrate selected. It is also possible to intensively process a stiff-coated base substrate in an all-over format thereby removing most of the surface dimension to produce a completely different material from the original. The laser devore effects are particularly striking on leather and suede-base substrates. Dependent upon textile properties, these laser devore materials may be easily labelled, wash and wear.

CONCLUSION

It is an irrefutable fact that devore is popular with fashion designers, it is also popular with the textile buying public. Devore, as a fashion fad, resurfaces approximately every ten to fifteen years; would it prove so popular next time it is reinvented if more information pertaining to the potentially hazardous environmental consequences of its manufacture were public knowledge?

Will this cyclical time span be sufficient to realise the creation of base substrates and dyes completely compatible with digital printing technology

The particular laser system currently used for processing is not capable of printing colour, but, with the advent and subsequent fast development of digital printing systems, it is only a matter of time before some enterprising agency combines the two technologies and a cleaner more efficient method of producing a full devore competitive product is realised, making it feasible to print without pre-coating base substrates and abandon the process of steam fixing inks.

Digital print technology and textile laser technology are still in their infancy, but the future looks promising for collaborative digital technology ventures, these in turn should revolutionise textile education, blurring the boundaries between textiles disciplines and producing multi-skilled practitioners. And, perhaps a new genre of devore.

CELLULOSE - PROTEIN TEXTILES:
UTILISATION OF SERICIN IN TEXTILE FINISHING

A. Kongdee and T. Bechtold
Christian-Doppler Laboratory for Textile and Fibre Chemistry in Cellulosics,
Institute of Textile Chemistry and Textile Physics, Leopold-Franzens University
Innsbruck, A-6850 Dornbirn, Austria

ABSTRACT

Sericin was used for cotton textile finishing in pad-dry-cure process. By means of Fourier Transform Infrared spectroscopy-Attenuated Total Reflectance (FTIR-ATR), fixation reactions of sericin on fabric surface were investigated. Estimation of sericin extent on fabric surface was performed using Supranolechtbordeaux B acid dye. The greater K/S values were observed for the samples treated with greater concentrations of sericin in finishing solutions. Variations in amount of sericin on fabric surface affected free formaldehyde content, crease recovery angle and water retention properties. With increase in concentration of sericin, free formaldehyde content and electrical resistivity of the samples decreased while crease recovery angle insignificantly increased with certain amount of sericin. The textile treated with sericin shows good wear properties and the finishing process is considered as a technique for protein fixation on the surface of cellulose textiles.

INTRODUCTION

Sericin has been currently exploited in many applications such as cosmetics, separation membrane, supplementary food and medical materials.[1] Modified fibres with sericin have been invented.[2-4] Its prevention in abrasive skin injuries and rashness has also been investigated.[3-4] N,N'-dimethylol-4,5-dihydroxyethylene-urea has been used as crosslinking agent for cotton fabrics in pad-dry-cure process[5-6], and the correlations between physical properties and dye ability of treated fabrics, and sericin content in finishing solution have been reported. However, permanent fixation of sericin on fibre surface has not been yet mentioned.

In this investigation, sericin was used to modify cotton fabric surface. The existence of sericin on fabric surface was analysed using FTIR-ATR. The treated fabrics were dyed with Supranolechtbordeaux B acid dye, amount of sericin on fabric surface were evaluated using the K/S function. Changes in free formaldehyde, crease recovery angle and electrical resistivity of the samples treated with various amount of sericin were evaluated.

EXPERIMENTAL

Preparation of treated fabrics

Finishing solutions containing chemicals as shown in Table 1 were prepared. 140x30 cm of woven twill, mercerised, and bleached cotton fabric was padded on a laboratory padder containing finishing solution, and then dried in laboratory dryer regarding to manufacturer recommendation. Each treated fabric was cut into two halves. One was retained without further treatment while another was washed with distilled water at 70°C for 30 min (liquor ratio of 1:50).

Table 1. Finishing solution recipes

Sample code	Component Sericin (g/l)	DMDHEU-based reagent (ml/l)	Magnesium chloride (g/l)	60% Acetic acid (ml/l)
UT	-	-	-	-
0 (blank)	-	40.0	10.0	1.0
2.5	2.5	40.0	10.0	1.0
5	5.0	40.0	10.0	1.0
10	10.0	40.0	10.0	1.0
25	25.0	40.0	10.0	1.0
50	50.0	40.0	10.0	1.0

Characterisation of fabric surface

FTIR-ATR was used to characterise surface of washed-treated fabrics and sericin powder. DMDHEU-based reagent was padded on KBr disc and characterised using FTIR.

Estimation of sericin content on fabric surface

Treated fabrics were dyed with a 2.5% (owf) solution of Supranolechtbordeaux B acid dye at a liquor ratio of 1:50, at pH 3.1, for 30 min at 60°C. After drying at room temperature, reflectance of dyed fabrics was measured at 541 nm on a Pye Unicam SP 8-100 double beam spectrophotometer with diffuse reflectance sphere 0°/d. Changes in sericin content on fabric surface were evaluated using K/S_{corr} values as calculated from equation 1.

$$\left(\frac{K}{S}\right)_{corr} = \left(\frac{K}{S}\right)_{sample} - \left(\frac{K}{S}\right)_{blank} \qquad (1)$$

Determination of free formaldehyde content

Content of free formaldehyde from freshly treated samples was determined according to LAW112.[12] 1 g sample was extracted with 100 ml of distilled water in closed bottle at 40°C for 1 hr. The extract was then filtered. 5 ml of aliquot was added to 5 ml of reagent prepared from 15% (W/V) ammonium acetate, 0.2% (V/V) acetylacetone and 0.3% (V/V) glacial acetic acid, and the mixed solution was kept at 40°C for 30 min. Absorbances of solutions were then spectrometrically measured at 412 nm. Formaldehyde content was obtained using a calibration curve constructed with defined amounts of formaldehyde.

Determination of crease recovery angle

Crease recovery angle (CRA) of the samples were performed according to DIN 53890 at 5 and 30 min. CRA of specimens in warp and fill directions was measured separately, CRA° (W+F) of specimens was reported.

Determination of electrical resistivity

Electrical resistivity of treated fabrics was measured using a Teraohmmeter (Siemens, 7 KA 1100). The samples were placed between two plate electrodes of 11 cm diameter, electrical resistivities of the samples were measured at a constant voltage of 100 Volt.

A set of identical determinations was conducted on untreated sample, which was treated as control. All samples were conditioned in a standard atmosphere of 65% RH at 20°C for a minimum of 48 hr prior to determination.

RESULTS AND DISCUSSION

Fixation of sericin on fabric surface

Figures 1 and 2 show FTIR-ATR absorbance spectra of treated cotton fabrics and finishing agents (DMDHEU-based reagent and sericin) respectively. When cotton fabric was treated with DMDHEU-based reagent alone, a peak appears at 1706 cm^{-1} as seen in Figure 1 (a). This peak corresponds to ν(C=O) in DMDHEU-based reagent as compared with FTIR spectrum of DMDHEU-based reagent in Fig. 2. These peaks show stable intensity shifts to 1695 cm^{-1} in spectra of cotton fabric treated with DMDHEU-based reagent and various amount of sericin because of reactions between DMDHEU-based reagent and sericin. In Figures 1 (c), (d) and (e), additional absorption bands appear at 1745, 1652, 1540-1520, 1457 cm^{-1}. These bands correspond to ν(C=O) of generated ester group from crosslinked reaction, and ν(C=O), δ(N-H) and ν(C-N) in sericin, respectively. From FTIR results, it was proved that DMDHEU crosslinked cellulose and sericin chains in the pad-dry-cure process. Besides the cellulose crosslinking operation, fixation of sericin occurred by chemical bonding to cellulose.

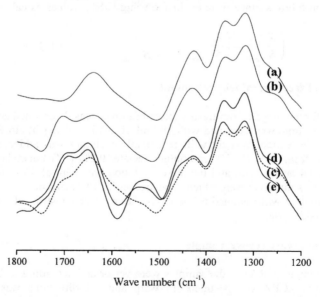

Figure 1. FTIR-ATR absorbance spectra of untreated fabric (a), fabrics treated with DMDHEU-based reagent alone (b) and with combination of DMDHEU-based reagent and sericin 10.0 (c), 25.0 (d), 50.0 g/l (e).

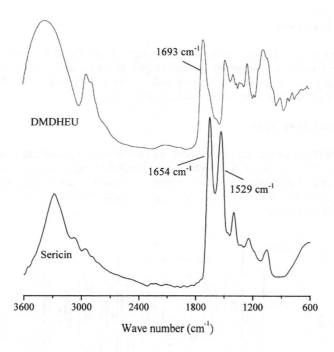

Wave number (cm^{-1})

Figure 2. FTIR absorbance spectra of DMDHEU-based reagent (on KBr pellet) and sericin powder (FTIR-ATR).

Content of sericin on fabric surface

Table 2 shows K/S values of dyed samples. Samples treated with the DMDHEU-based reagent alone lowered K/S from the value of untreated cotton. Increase in concentration of sericin in finishing solution is assumed to relate to greater sericin content on fabric surface. Therefore, increase in K/S$_{corr}$ with sericin content was observed as shown in Figure 3. K/S$_{corr}$ values of unwashed samples were comparatively less than those of washed due to hydrolysis of sericin during dyeing.

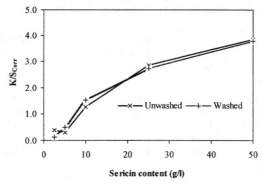

Figure 3. Relationship between K/S$_{corr}$ of dyed samples and concentration of sericin in finishing solution.

Formaldehyde content

Table 2 shows formaldehyde content in freshly treated samples. Dramatic lowering in formaldehyde content was observed for the samples treated with greater concentrations of sericin. Very low amounts of formaldehyde at 40 and 50 ppm are found for the samples treated with 25 and 50 g/l of sericin, respectively.

Crease recovery properties

CRA (Warp+Fill) of unwashed and washed samples were comparable as seen in Figure 4. The values of the samples treated with DMDHEU-based reagent dramatically increased compared to untreated control. Increase in sericin concentration on the sample surface affected CRA insignificantly. Although a slight decrease was observed when sericin >25 g/l was used.

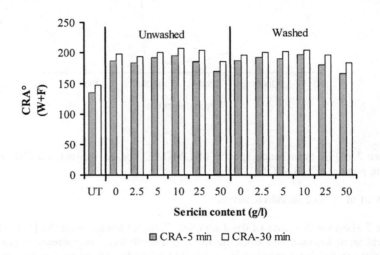

Figure 4. Crease recovery angle (Warp+Fill) against sericin content in finishing solution.

Table 2. Properties of unwashed (UW) and washed (W) samples

Sample code	K/S		Free formaldehyde (ppm)	Electrical resistivity (×10^7 Ohm)	
	UW	W		UW	W
UT	1.26	-	-	123.3	-
0	1.08	0.75	161	20.8	1112.3
2.5	1.47	0.89	123	14.0	1300.0
5	1.38	1.25	95	15.0	1166.7
10	2.33	2.27	82	17.2	1296.7
25	3.81	3.60	40	6.03	95.3
50	4.86	4.60	50	7.7	80.0

Electrical resistivity

Electrical resistivity of untreated sample was intermediate between unwashed and washed samples. The values of unwashed samples were much less than those of washed samples as the presence of conductive hydroscopic substance; magnesium chloride. In comparison with washed samples, electrical resistivities remarkedly decrease with greater amounts of sericin present on fabric surface.

CONCLUSION

Sericin was successfully crosslinked on fabric surface using a common finishing agent; a DMDHEU-based agent. Sericin was present on fabric surfaces as there were crosslinking reactions between DMDHEU, sericin and cellulose occurring during pad-dry-cure process, which was analysed using FTIR-ATR. Elevation in amounts of sericin on fabric surface lowered formaldehyde content and changed fabric properties. Electrical resistivity of treated samples decreased with increasing sericin content while crease recovery angle only slightly changed with certain amounts of sericin.

REFERENCES

1 Y -Q Zhang, 'Applications of natural silk protein in biomaterials', *Biotechnology Advances*, 2002 **20** 91-100.

2 P Jin, T Igarashi and T Hori, 'Application of silk sericin for finishing of polyester and nylon fabrics', *Sen'I Kogyo Kenkyu Kyokai Hokoku*, 1993 **3** 44-49.

3 U Yoshiharu, S Katsumi and T Kingo, Hydrophilic staple fibres, their nonwoven textiles free from skin irritation, and manufacture of nonwoven textiles, Japan Patent, Pat No 2003201677, July 2003.

4 S Zhaorigetu, N Yanaka, M Sasaki, H Watanabe and N Kato, 'Inhibitory effects of silk protein, sericin on UVB-induced acute damage and tumor promotion by oxidative stress in the skin of hairless mouse', *J of Photochemistry and Photobiology B: Biology*, 2003 **71** 11-17.

5 Y Kawahara, M Shioya and A Takaku, 'Effects of non-formaldehyde finishing process on dyeing and mechanical properties of cotton fabrics', *American Dyestuff Reporter*, 1996 **85**(9) 88-91.

6 Y Kawahara and M Shioya, 'Physical properties of chemically treated cotton Fabrics', *American Dyestuff Reporter*, 1997 **86**(11) 51-56.